D1116670

Turnaround, Shutdown and Outage Management

To my wife Liz
First, last and always,
to my sons
Julian and Craig
our hope for the future,
and to my grandsons
Jay, Callum and Adam,
the fulfilment of that hope

Turnaround, Shutdown and Outage Management

Effective Planning and Step-by-Step
Execution of Planned Maintenance
Operations

Tom Lenahan

ELSEVIER

AMSTERDAM • BOSTON • HEIDELBERG • LONDON • NEW YORK • OXFORD
PARIS • SAN DIEGO • SAN FRANCISCO • SINGAPORE • SYDNEY • TOKYO

Butterworth-Heinemann is an imprint of Elsevier

Butterworth-Heinemann is an imprint of Elsevier
Linacre House, Jordan Hill, Oxford OX2 8DP
30 Corporate Drive, Suite 400, Burlington, MA 01803

First edition 2006

British Library Cataloguing in Publication Data
A catalogue record for this book is available from the British Library

Library of Congress Cataloguing in Publication Data
A catalogue record for this book is available from the Library of Congress

ISBN-13: 978-0-7506-6787-6
ISBN-10: 0-7506-6787-7

For information on all Butterworth-Heinemann publications
visit our web site at http://books.elsevier.com

Printed and bound in the UK

06 07 08 09 10   10 9 8 7 6 5 4 3 2 1

Contents

Preface

Part one of this book describes the *process* of managing Turnarounds. It defines the rational, chronological steps, from initiation to termination. In the Preface to the earlier edition I wrote the following words about the book:

> *'I am not arrogant enough nor yet naive enough to believe that it is definitive, but it is a start, a foundation, if you like, upon which others can build.'*

At the time I wrote them I didn't suspect how true those words were. I suspected even less how much I still had to learn about Turnarounds and the people who do them and I had no suspicion at all about how many of my assumptions I would have to overturn in order to make the paradigm shift from someone on the inside looking out (a Turnaround Manager working for one large company with a single approach to Turnarounds) to someone on the outside looking in (as a Turnaround Consultant working with many companies each with a different approach to Turnarounds and, equally important, a different operating context).

The root of my paradigm shift was the conversion from the belief that:

> *There is one right way to do Turnarounds,*

to a much deeper level of understanding that:

> *There are basic principles that should be applied to effectively perform Turnarounds but the method of application will differ from company to company.*

I am still learning. Almost six years after the earlier edition of this book was published, the quest for excellence goes on. Each time I am called upon to audit, train, coach or consult with a client I invariably learn something that allows me to further hone the focus of the model of excellence.

Another welcome change that has occurred since the earlier edition was published is that there are now others (specifically Joel Levitt and Michael V. Brown) who have contributed books to the field and increased the general understanding of how these complex enterprises are planned and managed. I salute them.

Finally – looking to the future – the work I have done over the last six years has led me to the conclusion that the next stage in the development of Turnaround methodology, the way forward if you like, will be the application of systems principles to Maintenance in general and Turnarounds in particular – the exploration of the relationships between the various elements of Turnarounds to uncover the complexity, the interconnectivity and the resultant emergent properties that are generated. Only by understanding these can we truly understand Turnarounds.

Tom Lenahan
tomlenahan@hotmail.com

Acknowledgements

This book is the culmination of a ten-year quest to distil basic principles out of the rich and varied practice of Turnaround management. Many people have, over the years, contributed concepts, insights and suggestions which have added to the sum of my own knowledge and served to clarify the central theme of this book, namely that Turnaround management is a rational, coherent, process.

Special thanks first to three people: to Professor Anthony Kelly, Central Queensland University (and Honorary Fellow, Manchester University School of Engineering), who saw enough merit in my work to suggest that it was worth publication, to Simon Smith, who contributed time, materials and the introduction to Chapter 1, and to my son and colleague Julian, who organized the material for the initial draft of the manuscript and added some insights of his own.

Many thanks also to John Harris, Manchester University School of Engineering, who edited the complete text of the earlier edition, to Jonathan Simpson of Butterworth-Heinemann, and to the following colleagues for their part in the quest:

Earlier edition

Barry Stirling of ABB	John Billington of Tioxide
Eric Scott	Ian Adams of Foster Wheeler
Chris Greaves of ABB	N. Sankara Narayanan of Hofincons India
Steve Waugh of ICI	Claire Gulliver of ICI Paints

New edition

The faculty and MSc students at Manchester University who 'tried out' the Model of Excellence for me.

Part 1

The Process

1

Turnaround overview: context and strategy

Introduction

The Turnaround in a business context

It is perhaps an obvious statement to make that, given the complexity of the technology currently employed in the design and building of manufacturing plant, the limitations of the technology used to maintain it and the ever-increasing problem of its ageing, there will always be a need for maintenance work on the physical assets of the manufacturing activity. Furthermore, in an intensely competitive global market, characterized by increasing scales of production, the effective planning and management of that maintenance activity is coming to be seen as an ever more critical business process – one which is capable of differentiating the excellent performers from those who are merely capable, and from those who are less than capable.

It is also the case that with large scale assets there will be maintenance work which can only be carried out when the plant has been taken off line, and made safe for the performance of such work (which includes, but is not limited to, critical inspection, equipment overhaul, repairs and plant modification).

The case for avoiding such interruptions to production, particularly in continuous processing plants, is well made by those responsible for production and for the profitability of the company. As a consequence, a more serious and focused effort has recently gone into the design of technologies capable of monitoring and maintaining plants on line, minimizing costly outages with all their attendant risks to safety, reliability, business and the environment.

While the quest for production without regular plant shutdown goes on and the goal remains tantalizingly out of reach, there will remain the need – world wide and for the foreseeable future – to organize and carry out significant maintenance activity in the form of plant Turnarounds on much of the existing industrial asset base. The purpose of this book is to explain, in a structured and logical manner, an approach for achieving this with minimum of risk to the enterprise.

The approach has been developed over many years by Turnaround practitioners who have managed such events, large and small, on a wide variety of assets in Europe, the USA, Africa and Asia Pacific. These practitioners have generated, at both strategic and tactical levels, a corpus of principles, routines and processes that have proved to be a sound basis for managing these complex, hazardous and time-driven activities.

While it is clear that there have been major benefits in sectors such as oil and gas, petrochemicals, chemicals and utilities, the methodology will commend itself to other areas where the degree of sophistication may be less but where the business benefits may still be considerable, i.e. where large losses may be incurred by unnecessary planned downtimes.

It is interesting to conjecture why Turnarounds have not hitherto received this level of attention, given their significant impact on many companies' performances around the world. The answer may lie in the fact that most companies have a history of tolerating higher than necessary downtime, with outages becoming even more frequent as their plants get older, and perhaps, even now, accepting an annual shutdown as a 'necessary evil'.

With so much revenue at stake it is probably not too surprising that it has been the process sector that has led the way out of this frame of mind by constantly striving to lengthen the intervals between Turnarounds – from the traditional twelve or twenty four months to four, five and, in some cases, as much as eight years. This strategy has of course necessitated changes in plant operating standards and inspection techniques. In the UK this has been helped by the recent changes in the Pressurized Systems Regulations which have turned from the traditional prescriptive approach to something more flexible which, while allowing the companies that operate plants to use their technical expertise when setting intervals for inspection, nevertheless places the responsibility for the safety of the plant more squarely on the shoulders of those same operating companies. The corollary to this is that the planning and preparation for Turnarounds has to be carried out ever more carefully, aligning capital programmes, scrutinizing and challenging work scope, assessing plant deterioration and its likely impact on reliability, planning stocks and safeguarding supplies to customers, and partnering with major engineering contractors specializing in plant overhaul (who are now generally the only available resource capable of carrying out the scale of activity required).

It is the potentially daunting nature of these activities and the long lead times required to manage them effectively that has given rise to this book. The following chapters therefore seek to explore firstly, the strategic issues surrounding the key questions – such as the fundamental need for Turnarounds and how they will be managed – and secondly, the more tactical issues of how to plan, prepare and execute these events.

Reliability, the fundamental driver

In order to be profitable, a company needs consistent means of production delivered by reliable operating plant. There are various definitions of reliability and many company mission statements combine the words reliability, capability, utilization and the like in seeking to capture what it is they require from those who operate and maintain the plant. The general aim of this book is to strip techniques and processes down to their simplest form to lay bare what chemists used to call the 'active' ingredients – those which make the difference. In keeping with this principle the simple definition of reliable plant, for the purposes of this book, is

> *'a plant that is available when required and capable of performing to designed specification economically and safely for the life of the plant'*

Again, in keeping with the purpose of this book, maintenance, a physical expression of asset management, is defined as

> *'the sum of activities performed to protect the reliability of the plant',*

activities which will therefore help provide a consistent means of production and help generate profit.

Finally, a Turnaround is defined as

> *'an engineering event during which new plant is installed, existing plant overhauled and redundant plant removed'*

Because the Turnaround is a significant maintenance and engineering event a direct connection can be drawn between its successful accomplishment and the profitability of the company. There are a number of facets to this connection and a brief exploration of each will serve to set the Turnaround activity in a business context.

Cost of the event

In this context, a Turnaround impacts the business in a particular way. It is financed from company profits and, because it is typically an expensive event, it can have a major impact on those profits in the year in which it is performed. The cost of Turnarounds performed at intervals greater than one year may be spread over a number of years to lessen the apparent impact on one year's profit, but the loss of profit remains the same. Determining the total cost of a Turnaround is, at present, a subjective exercise because differences in culture from company to company mean that different factors and, in some cases, a different logic is used to determine it. For instance, should the profit lost because a plant is not producing for the period of the Turnaround be considered as part of the Turnaround cost? If not, why not? It is only when the total true cost of the event is known that its real impact on the business can be assessed.

Drain on resources

The Turnaround is a drain on company resources and often diverts personnel from other important work. In most cases the event requires many more people than are normally employed on the plant and external resources need to be brought in.

Hazard to plant reliability

The Turnaround is a potential hazard to plant reliability. Paradoxically, although it forms a major part of the maintenance strategy – the purpose of which is to protect the reliability of the plant – it can actually diminish or destroy reliability if not properly planned, prepared and executed, due to poor decisions by managers and engineers, bad workmanship, the use of incorrect materials, or because of damage done to the plant while it is being shut down, overhauled or re-started.

Potential safety hazard

The Turnaround increases the potential for harm to people, property and the environment. Compare the normal routine on the plant to the situation which exists during the Turnaround. Normal routine is characterized by the presence of a (relatively) small, experienced team performing familiar tasks using (it is to be hoped) well defined procedures in order to make products which will be sold at a profit. The Turnaround reverses almost every characteristic of normal routine; the plant is shut down, taken apart and worked on by a large number of strangers using unfamiliar and inherently hazardous procedures and equipment. Under these circumstances the potential for accidents rises almost exponentially and must be matched by a safe system of working which minimizes the probability of loss. This adds to the cost of the Turnaround and may even be dwarfed by the consequences to the company of actual loss, in terms of monetary cost or loss of reputation.

Risk of overspend and overrun

Finally, because of the technical uncertainty inherent in the Turnaround due to the possible occurrence of unforeseen problems, there is an ever present risk of the cost estimate being exceeded, the duration of the event being extended, or both. The prudent company will therefore take this into consideration by building time and cost contingencies into the Turnaround plan (which adds to the cost) – but even these may be exceeded. The resultant extra costs and loss of revenue may be substantial.

Strategy

Taking all of the above into consideration, as well as other factors which will be dealt with in this section, it is incumbent upon the senior management of the company to create a business strategy for managing Turnarounds. The basic objective of this strategy – and this may seem an unusual perspective in a book dedicated to the management of such events – should be to eliminate maintenance Turnarounds altogether. Thereafter, if it should prove that the event is absolutely necessary, senior management should ensure that the Turnaround is aligned with maintenance objectives, production requirements and business goals.

Steering group concept

A company which is committed to getting the best value from its Turnaround programme can realize this commitment in a concrete way. The most senior manager in the company should form a *steering group* consisting of senior managers who take responsibility for the long-term strategy for Turnarounds. The committee will meet at regular intervals throughout the year to review current performance and formulate high-level strategy for the management of Turnaround events and the long-term Turnaround programme.

Long-term view

On any plant which has been operating for a number of years some Turnarounds will already have been performed, and in every plant covered by the imperatives of this book there will be future Turnarounds. Rather than being seen as individual events separated by a number of months or years, Turnarounds are best viewed by senior management as a sequence of linked activities (with many identical features) which are performed as part of the continuing process of asset management and are tied together, in the long term, by inclusion in senior management's business strategy.

Consideration of past practice, present intentions and future requirements gives a well-rounded perspective and prompts the types of strategic considerations which follow. Detailed questions for the following elements are contained in Figure 1.1.

Turnaround philosophy

To be truly aligned to overall business strategy, the evolution of asset management should be driven by the constant search for ways to change from the current vogue of preventive maintenance – which is driven by technical considerations – to a philosophy of maintenance prevention, which is driven by business needs. While it may not be possible to achieve zero maintenance, the only way to find out how nearly we can approach this utopian ideal is to question every maintenance practice to determine if it can be eliminated (by addressing the cause that generated the need for it). This being the case, it is logical to examine the largest maintenance initiatives, namely Turnarounds, first.

Questions to be posed by senior management at an early date to minimize the impact of the turnaround on business performance

Turnaround philosophy
Can we avoid doing turnarounds?
Defer or delay turnarounds?
Do them less frequently?
Do them simpler or cheaper?
Change methodology or technology?
Out-source significant work packages?
Form an alliance to out-source work?

Financial concerns
Can we afford this turnaround?
Does early estimated cost of event compare with allocated budget?
What percentage of maintenance budget does this cost represent?
What is the projected loss of profit due to the event?

Current performance
Is current level of plant maintenance achieving desired reliability?
Is current plant performance at an acceptable level?
Is the product quality currently conforming to specification?
Is the cost of production increasing, decreasing or stable. Is it acceptable?

Plant turnaround history
Does the event work scope seem to deal with known plant problems?
Have previous turnarounds had any negative effects on the plant?
What percentage of work done on previous events was unpredicted?
What lessons have been learned and how will they be incorporated?

Turnaround objectives
Only do if absolutely necessary
Maximum time between events
Validated work scope!
Minimum duration?
Lowest cost?

Corporate knowledge
Do we have the capability to manage the turnaround programme?
How cost effective are we compared to the best in the field?
Do we have a strategy for reducing costs in the face of ageing plant and rising manpower and material costs?
Where can we get high level advice?

Project work scope
What percentage of the total work will projects represent?
Is the project production, safety or quality driven?
How much will each project cost in total and what value will it add to the business?
Can any project be cancelled?

Maintenance work scope
What is the business justification for doing the major items of work?
Is there any ritual work on the list?
Can we eliminate inspection work by using risk analysis?
What percentage of the work scope is bulk work?

Organization
What manpower is available in-house?
What are the competence levels?
Where do we get additional resource?
Who will manage the event and who will do the work?
What will be the design of the event organization?

Figure 1.1 Questions to be posed by senior management at an early date to minimize the impact of the Turnaround on business performance

The very first question (a negative answer to which would obviate the necessity for the succeeding questions) is

'is the Turnaround necessary at all?'

This question – the fundamental one – should be given the serious consideration its potential payback deserves. It should not be dismissed lightly. It is a powerful and difficult question because it forces many managers to abandon the comfort zone they have inhabited for so long, with regard to the subject of maintenance in general and of Turnarounds in particular.

Every maintenance task carried out costs money that is subtracted from the bottom line, the profit margin. At the front end of the business, disciplines such as procurement, production planning, sales and marketing are being honed to a fine edge by use of the latest technology, massive management input and constant re-evaluation, while at the back end disciplines such as maintenance (and especially Turnaround management) have undergone, in many companies, little more than cosmetic change in the past thirty years. A simple exercise to bring home the seriousness of this situation is to calculate the total cost (including the cost of plant-based personnel and the profit lost due to the plant being off line) of the last Turnaround performed by the company and divide it by the estimated daily profit of the company. The result will indicate the number of days that the company had to work, on full production, simply to finance the activity. To use a football analogy, it's no use scoring goals at one end of the park if you are scoring own goals at the other end.

A common response to the above is the statement that, in the real world you cannot avoid the necessity to maintain the plant (Turnarounds being, in part, simply large maintenance events). The crucial distinction in this statement is the definition of the term maintenance. It generates a number of questions, each of which is crucial to understanding the high cost of maintenance and Turnarounds. They are:

How many of the tasks on a Turnaround work list are generated by:

- Legal requirements imposed upon the company?
- Actual maintenance needs based on equipment manufacturers' recommendations and genuine requirements evolved by experience?
- Fixing chronic problems which have never been investigated to find the root cause?
- Ritual tasks which are simply unnecessary?
- Avoidable attrition by process fluids or product substances (corrosion, fouling etc.)?
- Shortcomings inherited from design, procurement, installation and commissioning?
- Inadequacies in the approved plant operation procedures?
- Incorrect operation of the plant?
- Insufficient environmental protection?
- Lack of competence on the part of those (at any level) who operate and maintain the plant?
- Insufficient preventive maintenance throughout the year?
- Unrealistic expectations by the plant owners (overloading the plant)?

All such tasks generate cost, but are they all necessary? If they are accepted unquestioningly as 'maintenance' tasks then the definition of 'Turnaround' must be expanded from the maintenance definition, i.e.

'the sum of activities performed to protect the reliability of the plant'

to

'the sum of activities performed to protect the reliability of the plant against all comers'

If, on the other hand, it is recognized that all of the tasks on the list can be investigated with a view to eliminating, or at least minimizing, them, then the first step has been taken towards challenging the necessity for Turnarounds. Also, in choosing to perform a Turnaround (no matter how complex and difficult the planning and execution of the event may be) the management may be sacrificing profit by choosing the easy option.

If, after the careful consideration outlined above, the management still believes it has no alternative but to perform the Turnaround then the same challenging questions should be addressed at the level of the single task, by the appointed planning team, to ensure that the desired outcome is achieved with the minimum input.

Financial concerns

Turnarounds often cost more than is realized, because many companies do not calculate the total cost, concerning themselves only with the direct cost of planning and executing the event. Unless every element of cost thus generated is taken into consideration, control is not possible. We can only affect what we perceive.

The fundamental financial concern should be to plan and execute the event as cost effectively as possible. Using 'acceptable cost' as the tool to examine all expenditure will sharpen management focus and reduce the myriad elements of the Turnaround to a single measurable and controllable unit – money.

If money is to be used as the measure then it is imperative that the scope of work be effectively defined at the earliest possible date. The work scope is the foundation upon which all other aspects of the Turnaround rest and will have a major influence in determining the final cost. The logic is simple; to exert control from the earliest possible date, senior management must have as accurate a cost estimate as possible – and this depends upon effective definition of the work scope.

An early cost estimate (even if only in 'ball park' terms) will indicate whether the budget for the event is sufficient. If not, senior management must choose between providing extra funds to cover the shortfall, eliminating elements of the work scope to bring the estimate back within budget, or finding some other strategy to optimize cost and work scope (the options are discussed in more detail in Chapter 6).

A second important consideration is how much of the overall maintenance budget should be allocated to the Turnaround. This will depend upon the age and nature of the plant, the company's maintenance philosophy and the scope of the work. (A significant amount of benchmarking of maintenance costs has been done by companies such as Salomon Bros in the USA, and IPA in the UK.)

A third consideration is the amount of profit lost

(a) while the plant is off line due to the Turnaround and
(b) on subsequent days when plant production rate or product quality may be below target.

The final consideration here, a cost element often overlooked or hidden, is the wage and salary bill for plant personnel during the Turnaround. It is argued by some managers that employees have to be paid whether the plant is operating or off line, and that the costs should therefore be allocated to production. The counter argument is that plant employees are paid to produce and if, during the Turnaround, they are not producing then their cost should be allocated to the event, which otherwise will appear to cost less than it actually did (and, conversely, the cost of *production* will appear to be greater than it *actually* is).

Current performance

The scope of the Turnaround work may be dictated by the current operating performance of the plant and the level and effectiveness of preventive maintenance carried out in the periods between Turnarounds. Rationalization and 'right-sizing' has led many companies to reduce their maintenance personnel to a minimum. One negative side-effect of this is that some preventive (and even some corrective) work does not get done. There is a great temptation to dump this into the Turnaround work scope, to relieve the pressure on the maintenance crew, but it must be borne in mind that (due to higher management costs and loss of profit) it costs significantly more to perform a task during the Turnaround than during normal operation, the work scope may be unnecessarily encumbered, resources may be diverted from more critical work and the duration of the event may be extended. The golden rule is:

The only tasks allowed on to the Turnaround work list are those which cannot be done at any other time, unless there is an overriding reason (e.g. maintainability, hazard etc.).

The Turnaround should not be used to mask the shortcomings of the company's maintenance philosophy.

Regarding the work list which is generated by the current performance of the plant, care should be taken to ensure that the remedial work requested will actually address the problem being experienced.

Plant Turnaround history

Turnarounds are performed to protect the reliability of the plant. The plant should perform as well or better after the event than it did before it. An examination of past Turnaround performance (where this is possible) and of subsequent plant performance will indicate whether the Turnarounds have provided the protection expected. If not, the situation must be reassessed to find out why and a new rationale for Turnarounds developed. Past events should also be analysed to ascertain the ratio of emergent work (which only arises after the execution phase has begun) to planned work and the extent to which emergent work increased the planned expenditure (the norm is between 5 and 10 per cent, the worst case in the author's experience being in excess of 45 per cent). If past levels of emergent work are unacceptable, the senior management should ensure that plant personnel address the problem and improve the quality of technical specification in Turnaround work requests.

Corporate knowledge

Turnarounds are a sequence of infrequent events in the life of the operating plant and they require management expertise significantly different from that required to operate and maintain the plant under normal circumstances. Ineffective management of the event can add more to its cost than any other factor.

Senior management must honestly assess whether their own people have the capability to manage and control the event. Even if technical capability exists in individuals, the cost effectiveness of internally managing the event should be analysed from the perspective of forming the technically competent individuals into a team capable of managing its total conduct. Again, a good start is to benchmark against the best in the field.

The expectation of senior management should be for the cost of each subsequent Turnaround to be reduced – by the application of greater knowledge, superior technology and better organizational ability, even in the face of ageing plant and rising costs. This requires expert, professional, event management.

Project work scope

In terms of an operating plant, a project is a discrete package of work performed for the sole purpose of improving plant performance in some way (otherwise it is surely a waste of time, effort and money). Demands for projects are generated from various sources – statutory safety requirements, production or quality improvement programmes and the like. Senior management should evaluate each proposed project in terms of necessity, desirability and payback. The obvious questions are:

- Is the project work really necessary and vital to the safety, quality or production efficiency of the plant?
- Is it desirable at this time in the life of the plant; is it too late or too soon?

What will the payback be in terms of increased profit or reduced costs? The author once attended a de-brief meeting at which the business area manager stated that

> 'During the last few Turnarounds the company has spent several million pounds on projects. The payback to date has been nil – the plant is not performing any better than it did before all this work was done'

It is not sufficient for projects to be technically desirable, they must also be justified from a business point of view.

It is a rare Turnaround that consists solely of maintenance tasks. More typically, the event will involve a mix of maintenance and project work. The ratio of one to the other will be determined by the current needs of the plant and the expectations of the management. The ratio is important because an event comprising 90 per cent maintenance and 10 per cent project would be managed differently from an event with the opposite proportions. Also an event involving a large number of small to medium sized projects would be managed differently from one involving only one very large project.

Whereas the Turnaround manager is responsible for the planning and execution of all maintenance tasks, projects are planned and, in many cases, resourced by a project manager or engineer. This means that on a large event there could be a number of project teams who execute their work independently of the maintenance work. However, for the event to be coherent all of the work – maintenance and projects – should be integrated into a single plan. The task of senior management is to decide how this is to be done and how responsibility and authority are to be apportioned.

Maintenance work scope

Maintenance work can be divided into three approximate categories:

- *major tasks*, such as the overhaul of a large machine or the re-traying of a large distillation column, which require engineering input;
- *minor tasks*, such as the cleaning and inspection of a small heat exchanger, requiring only the specification of the work that is to be done;
- *bulkwork*, the overhaul of a large number of small items such as valves and small pumps, which simply needs to be scheduled.

When evaluating the early work scope, the senior management should challenge the inclusion of each major task to ensure there is a sound business justification for doing it in the particular Turnaround or whether it can be eliminated altogether or deferred to a future event. Special consideration should be given to those tasks which will determine the critical path, and therefore the duration, of the event. This is the exercise of risk management and must be based on effective risk analysis.

Risk is defined as:

the numerical probability of loss occurring.

The two important questions to be asked are:

1. What is the specific nature of the harm (to people, production, property or the environment)?
2. What is an *acceptable* numerical probability?

There are a number of systems which may provide an objective answer to the latter question, by providing tables of probability and predicted failure rates, but in the end it comes down to this: what does the management of the company, at the time, believe to be an acceptable level of risk, given the set of unique circumstances which surround the operation of the plant? The answer depends upon the experience of the managers, the assumptions they use to frame their questions and the confidence they have in their own abilities.

In the context of the preceding paragraph, bulkwork may seem insignificant but the author would venture to suggest that almost every experienced Turnaround manager has had at least one example of an event in which valuable programme time was lost, not because of any problem with any of the major tasks (which were probably planned down to the last detail) but because of the bulkwork which, due to its large numbers and wide dispersion, in some measure got out of control. The senior management should examine the strategy for scheduling and managing bulkwork and should assess its effectiveness.

Organization

The design of the Turnaround organization will depend upon a number of questions concerning the planning, execution and management of the event which can only be answered by the senior management. There is a whole spectrum of types of organization which can be designed for the event (see Chapters 5 and 7). To illustrate this we will concern ourselves here only with the options which lie at either end of the spectrum. At one end is the option for the company to manage the event using their own managers and engineers. This begs the two questions:

- Are there enough people within the company to fill the required roles?
- Is the level of competence of each individual appropriate to the role required?

The second question may seem all too obvious but experience has shown that people are often assigned to jobs on the basis of their availability, with little regard to the question of their competence.

At the other end of the spectrum is the option to outsource the total Turnaround package to a contractor whose business is the planning and execution of Turnarounds. The main questions here are:

1. What is the contractor's track record – especially on financial claims against the client for variations to contract?
2. Can the management feel comfortable about putting their plant into the hands of a contractor without putting in place a control team to monitor performance? (A question which can become circular, because someone in the company is bound to ask whether – if the company has to field a team to monitor the contractor – it ought not use that team to manage the event directly and cut the contractor management out?)

Clearly the formation of the event organization requires much thought.

An engineering perspective

In the context of engineering, Turnaround management may well be unique. Other functions – such as projects, construction and production – are concerned with creating something new, whether it be programmes, plants or products. Turnaround management, on the other hand, is mainly concerned with the replacement, repair or refurbishment of plant which is worn, damaged or malfunctioning in some way. On other types of project, work can be planned in the knowledge that if their plan *is* disturbed it will be by some external force. Turnaround planning is by its very nature uncertain.

The planning and preparation which goes into a Turnaround is done while the plant is on line. Faults may be hidden and may only emerge when the plant is shut down, opened up and inspected. This makes it difficult to predict the exact work scope, and so there is an inherent uncertainty regarding the costing, resource needs and duration of the event.

The psychology of Turnarounds

Another issue that adds to the complexity of Turnaround management is probably best termed the *psychology* of these events. Consider the differences, already discussed, between the normal routine of a plant and the situation which prevails during a Turnaround from the point of view of the operators of the plant. Their plant which, as far as they are concerned, was running perfectly well, will be shut down (*no production – no profit*) and literally pulled apart (*a messy, confusing, situation*) by a large number of people who do not normally work on the plant (*strangers*) – this after a period of time when all of their requirements have been ruthlessly challenged (*by these strangers*) and their technical records and operating procedures will have been subjected to close, external, scrutiny (*and possibly found wanting*). On the other side of the fence there is the team charged with managing, planning and executing the Turnaround. They are working hard in the plant team's best interest and are confused by the fact that the latter seem sometimes less than enthusiastic about their help, and often seem to be obstructing their best efforts.

The potential exists for friction between the teams. Management's task is to supervise the issue and ensure that all personnel are integrated into a single team – no mean feat. The integration can only be accomplished by briefing, training, negotiating, convincing and, above all, establishing good clean channels of communication. Failure to address these issues can result in the Turnaround manager being faced with messy human problems which can be far more difficult to overcome than any technical obstacle.

Integration

The Turnaround organization – the group of people who, under the overall direction of the Turnaround manager will plan, prepare and execute the event – is formed by integrating five different types of knowledge and experience. These are:

1. *Local*: provided by plant personnel who are familiar with the technical records, operating procedures, history, past performance and current problems of the particular plant.
2. *Work management*: provided by engineers and planners who are familiar with all aspects of the initiation, preparation, execution and termination of Turnarounds.
3. *Supervision*: may be provided by client or contractor personnel.
4. *Craft*: provided by (internal or external) maintenance personnel who are familiar with the tasks and activities required to overhaul plant items.
5. *Specialist*: knowledge and experience in discrete technical tasks provided by the vendors' representatives and other specialist sub-contractors.
6. *Management of discrete projects*: provided by experienced project managers or engineers.

The phases of a Turnaround

A Turnaround is normally considered to be an engineering event of relatively short duration, but it is only one segment of a cyclical process with four phases – initiation, preparation, execution and termination – each of which has its own specific set of critical issues and activities (see Figure 1.2). It is rightly referred to as a cycle because the initiation phase of the next Turnaround should follow on from the termination phase of the current one.

Phase 1: Initiation

The reason for the inclusion of this phase (most other approaches start with the planning and preparation phase) is that this book attempts to consider the *entire* Turnaround process from the moment some senior manager flags up the necessity to start considering the requirements for a forthcoming event (this could be two years or more before the event itself). It is therefore necessary to define, in detail, the strategic issues to be addressed and the activities required to move the process to the point where it can actually be planned and prepared. This phase is characterized by defining objectives, setting policy and appointing the necessary personnel to set up the preparation team and gather basic data.

Critical issues

- Reviewing past events
- Setting objectives
- Formulating policies
- Minimizing costs
- Balancing constraints
- Delegating authority
- Monitoring performance
- Flexibility of approach

- Leading a small team
- Translating policy into a defined project
- Defining contingencies
- Creating a plan
- Reviewing contracts
- Forecasting costs
- Setting targets
- Selecting people
- Formulating rules
- Delegating authority
- Monitoring performance
- Resolving issues

- Co-ordinating activities
- Controlling planned work
- Controlling emergent work
- Achieving duration
- Minimizing expenditure
- Monitoring safety
- Monitoring quality
- Resolving issues

- Analysing performance
- Reviewing work done
- Extracting learning
- Recommending changes
- Closing out report

Turnaround phases

Phase 1 – Initiation
Period during which turnaround parameters are defined, core personnel appointed and basic data organized. Can be spread over a period of months

Phase 2 – Preparation
3 to 18 month period (depends on the size of the event) during which a large quantity of data, technical and non-technical, is validated and transformed into a set of plans that will be used to execute the turnaround
    The vital element in this phase is the work list because it forms the foundation upon which all other elements are built i.e., safety, quality, costs, materials, equipment, resources, logistics and duration
    Preparation is characterized by close attention to detail and accurate calculation

Phase 3 – Execution
Normally 2 to 8 week period when planned work is carried out and monitored against the event schedule, duration, cost, quality and safety requirements
    The emphasis during this time is on effective control of work

Phase 4 – Termination
Normally 1 to 2 week period when the work is closed out and performance is reviewed

Critical activities

- Senior business manager forms a steering group
- Team appoints a turnaround manager as its agent
- Manager mobilizes and leads preparation team
- Preparation team collect and collate basic data
- Plant team issue work request forms and provide technical data, plant history and local knowledge
- Plant team issues the initial work list

- Preparation team and plant team challenge and validate the work list
- Turnaround manager freezes the work list
- Planning officer and preparation team –
  – prepare job specification packages
  – identify and plan pre-shutdown work
  – identify and procure long delivery items then procure all other necessary items and materials
  – define contractor work and place contracts
  – organize site logistics and create plot plan
- Preparation engineer works up major tasks
- Turnaround manager optimizes the following –
  – work schedule and resource profile
  – organization chart
  – cost estimate and expenditure control system
  – contractor list and project list
  – safe system of work and quality plan
  – briefing document and site rules
- Turnaround manager submits plans to steering group for discussion, decision and approval
- Preparation and plant teams brief all personnel

- Plant team shut plant down to a prearranged plan (product off) with support of execution team
- Execution team perform work to turnaround plan
- Control team define and cost emergent work
- Execution team complete all work
- Turnaround manager demobilizes execution team and mobilizes start-up team
- Plant team start up plant to a prearranged plan (product on) supported by the start-up team

- Start up team clean the site and remove equipment
- Plant manager inspects plant and accepts handover
- Turnaround manager demobilizes start-up team
- Plant and turnaround manager organize debriefs
- Turnaround manager produces final report

Figure 1.2 Turnaround overview

Phase 2: Preparation

This is the major phase of the process. A small team of people work over a long period of time to specify, schedule, resource and cost the large volume of tasks required to perform the event. It contains an element of uncertainty because it involves some prediction of unknown conditions (of plant items) which, in turn, can involve everything from informed technical assessment to 'fingers crossed' guesswork. This is because the subject of the planning and preparation, the internal condition of the plant equipment, is hidden from view. One way of dealing with the uncertainty is to conduct an *analysis of contingencies*, i.e. to answer, in this case, the following questions:

- What faults are likely?
- How long will it take to put them right?
- How much will it cost?
- What impact will it have on duration?

A contingency analysis will provide only approximate answers, but they are, in many circumstances, the only guides available.

The final act of preparation is to communicate the requirements of the Turnaround to every single person who will be involved at any level. This is accomplished by a series of briefings carried out by the plant team and the Turnaround team.

Phase 3: Execution

In this phase the planning and preparation are tested against reality. Was the planning accurate? Was the preparation adequate? Was the contingency analysis sufficient? Execution is characterized by the performance of a large volume of tasks by a large number of people of many skills and disciplines, in a limited space and at different levels simultaneously, under (sometimes severe) time and financial pressure. The effective control and co-ordination of work is of paramount importance.

Execution can be broken down into a number of sub-phases, as follows:

- shutting the plant down (removing inventory, decontaminating, cooling, isolating);
- opening the plant up (physical disconnection of items and removal of covers);
- inspecting the plant (visual and instrumental examination and report);
- installation of new items, overhaul of existing items, removal of redundant items;
- boxing the plant up (final inspection, replacement of covers and reconnection);
- plant testing (pressure tests, system tests, trip and alarm tests);
- starting the plant up (re-connecting services and re-introducing inventory);
- plant clean up and final inspection (removing all traces) of the Turnaround.

In reality, some of these activities will overlap, but on any given task the activities will occur in this sequence and, more importantly, the transition from one activity to another involves, in most cases, the transfer of responsibility for the progress of the job from one person to another. This situation highlights a fundamental characteristic of properly co-ordinated events, namely *single point responsibility* which demands that the person responsible at any point in the process must fulfil four requirements, viz:

1.   Check that the previous stage of the task has been properly completed before taking it over.
2.   Ensure that the current portion of the task is carried out to specification.
3.   Hand the task over to the next stage at the earliest possible time.
4.   Report the handover.

This ensures that, irrespective of the length or complexity of the task, there is always someone directly responsible for its quality and progress. It is the Turnaround manager who, in the final analysis, is responsible for co-ordinating the total event and who must ensure that single point responsibility is understood and operated by his teams.

Phase 4: Termination

There are two separate elements involved in terminating the event. The first is ensuring that the plant is handed back in a fit condition and the second is the de-briefing of every member of the Turnaround organization, the latter in order to capture the lessons to be learned from the event so that subsequent Turnarounds may be performed more effectively.

These four phases go to make up the Turnaround process and even this brief overview indicates the complexity of its management. To be effective, it must be initiated, prepared, executed and terminated using a rational methodology which will give repeatable results. The remaining chapters of this book detail such a methodology. Rather than simply describing the separate activities required to perform a Turnaround they are organized in as strict a chronological order as possible, in order to give a sense of the way in which the process unfolds over a period of months.

2

Initiating the Turnaround

Introduction

At some point in time the senior management of a company will make the decision to initiate the process which will eventually lead to a plant Turnaround. With a large and complex plant, this may be up to two years before the event; for a small plant, as little as three months. Whatever the time scale, there needs to be a period when strategic decisions are made, key personnel appointed and basic data gathered to allow the planning and preparation phase to proceed in a coherent manner.

The steering group (see Figure 2.1)

A Turnaround is a large, expensive and time-consuming undertaking. Even a small event is, in relation to normal operation, a large undertaking because, although the work scope may be small, all of the activities described here have to be performed. The event requires close planning and preparation and will, in all probability, involve every department in the company to some extent. It therefore requires serious consideration by the senior manager responsible for production to ensure that it is performed safely and cost effectively. Its complexity, however, puts it beyond the scope of any one individual and the senior manager therefore has to appoint a steering group (or policy team) the permanent members of which should be drawn from managers, executives and engineers who fulfil at least one of the following three criteria:

1. They are 'stake holders', i.e. are directly affected by the Turnaround.
2. They provide the money to pay for the Turnaround.
3. They have the authority to make decisions concerning the Turnaround.

A typical steering group consists of representatives from the following functions (job titles have not been used because different companies use different titles for the same function or the same title for different functions). The main functions are:

- Senior management
- Marketing
- Production

Figure 2.1 The steering group

- Maintenance
- Engineering
- Projects

   The steering group will also have the authority to call upon any other function to attend specific policy team meetings to provide particular expertise on an 'as-required' basis.

The purposes of the steering group (see Figure 2.2)

Once appointed, the policy team assume the responsibility of ensuring the success of the Turnaround. As will now be explained, they have six specific purposes which focus their activities and decisions.

(1) The provision of funds for the Turnaround

The group controls the budget and ultimately decides how much will be allocated to the Turnaround. The actual cost estimate derived from the approved work scope will determine how much money is required. If the budget exceeds the estimate then the required amount may be allocated. If however, the reverse is the case then the group can pursue one of two options, either raise more capital to fund the approved work scope and duration or, by some means, reduce the work scope and/or duration to suit the budget. The group must also decide whether a contingency fund is to be

Figure 2.2 The purposes of the steering group

set aside for unscheduled work which may emerge during the Turnaround – and, if so, how much.

(2) The balancing of Turnaround constraints

The Turnaround will take place within a business environment which includes, but is not limited to, business performance, market circumstances, customers, competitors, the local community, available technology, legal requirements, health and safety at work, the local community and even the weather. The constraints are often in conflict with each other and have to be balanced. For instance, from a business point of view the cheapest and easiest way to get rid of surplus inventory while shutting a chemical plant down would be to dump it down a drain but that is (in most, if not in all, countries) illegal. Another option would be to flare it off – which, under certain circumstances, might not be illegal but the noise and flame of the burn might upset local inhabitants and adversely affect the company's standing in the community. The principle operating here is that every decision may achieve a benefit but will also incur a consequence. Benefits and consequences must be balanced in order to reach the optimum outcome.

(3) The setting of objectives for the Turnaround

So that people know where to focus their efforts, any plan needs objectives. Turnaround objectives must be measurable and must deal specifically with quantity, quality, time, money and safety (QQTMS). These include, but are not limited to:

Quantity: How much work will be done or how many people will be employed?

Quality: What is the acceptable minimum level of workmanship?

Time: When will the event begin and how long will it take?

Money: What is the acceptable cost of the event?

Safety: What is the tolerable rate of accidents, incidents, emissions? Should be zero.

As with constraints, the first *four* objectives must be balanced, i.e. each objective must be set with reference to all the others and with due regard to current constraints. For instance, it is unrealistic to set a duration for the event of twenty-one days if the longest activity on the critical path is going to take twenty-three days; or if seven hundred men are required to achieve that duration and only five hundred are available.

The *fifth* objective, safety, is in a different class. The safety of people, property and environment must not be compromised to achieve other objectives. Ever!

(4) The formulation of policy to meet the objectives set

Policy is simply the course of action which is taken to achieve the objectives, and is applied to the same QQTMS elements. For example:

Quantity: A mixture of internal and external resources will be used because the former is insufficient to cover the event.

Quality: The ISO 9002 quality standard procedures will be adopted, including audits of all sub-contractors to ensure uniformity.

Time: Because of market demands, the window of opportunity is limited to twenty-two days (between 4th and 26th of September).

Money: A 40 per cent increase on the allocated budget will be borrowed from an external source to meet the increased work scope.

Safety: A professional safety manager will be engaged for the event because the company cannot field anyone with the appropriate experience.

The policy decisions taken at this level and at this time will affect every aspect of the Turnaround throughout its four phases.

(5) The monitoring of progress against objectives

As mentioned before, objectives must be measurable so that progress can be monitored and compared against some standard to ensure that all phases of the Turnaround are kept on track. Again, progress is measured against QQTMS in terms of quantities fulfilled to date, compliance with quality requirements, key dates achieved, current cost expenditure and daily accident/incident rates. The steering group must formulate a set of procedures to accomplish this (see example later in this section).

Where a written set of procedures exists, formal audits may be used as a monitoring device. Typically, a large event might have a sequence of six audits as will now be described. (Smaller events should be subjected to at least one preparation, one execution and one termination audit, while very large ones might benefit from several extra audits, the guiding principle being to perform the minimum number of audits required to ensure that preparation, execution and termination of the event are adequately performed).

Audit 1 Strategy audit

Carried out at the earliest possible date. It examines the strategy formulated by the senior management and evaluates it against the stated business aims of the Turnaround.

Audit 2 System audit

Examines whether the systems that have already been put in place – or are being considered for later inclusion in the programme – will be adequate for effective implementation of the steering group's strategy.

Audit 3 Compliance audit

Examines planning and preparation, in all their aspects. Assesses whether they have been adequate to achieve the objectives of the Turnaround and, in particular, whether the personnel involved are complying with the systems.

Audit 4 Preparedness audit

Carried out two to four weeks before the start of the event, it examines the state of readiness of the Turnaround organization and the systems that will be used during execution to retain control of the event.

Audit 5 Execution audit

As its name implies, is carried out during the execution phase and examines the control system to assess whether the Turnaround manager and his organization are in control of the event. It also examines the actual work methods being employed and evaluates them for compliance against the planned procedures.

Audit 6 Final audit

Carried out during the termination phase. Examines the effectiveness of the de-briefing system in capturing the lessons to be learned and also of the system for ensuring that the information is recorded in such a way that it will be available for future events.

It is up to the steering group to decide if they wish none, some or all of the audits to be carried out and – for those that are to be performed – whether they should be carried out by independent auditors or by company personnel. The timing of the preparation phase audits is dependent upon the length of the preparation period.

(6) The modification, if necessary, of objectives or policy

Because a Turnaround is a complex process performed over a long period, the steering group needs to be flexible in its approach, able to modify its objectives or change its policies so that it can readily deal with changes (which may be radical) in process or environmental constraints. For example, the cost of borrowing the money to increase the budget by 40 per cent may suddenly be subject to a sharp rise in interest rate. Or new safety legislation may force an increase in the duration of the event.

The steering group needs to deal with these disturbances as they occur. The most acute disturbance is one that occurs *during* the actual event. A fault may be found in a critical piece of equipment and there may be no option but to carry out an extensive repair which will be costly and will increase the duration of the event. The problem here is one of *force majeure* because there is nothing that can be done to avoid it. In addition, the group may have very little time to react. The ability of the group to act competently will determine the success or failure of every other aspect of the Turnaround process, hence the strict criteria for the selection of personnel to serve on the group.

The methodology

To ensure objectives are met, the steering group should adopt a rational methodology comprising the following basic elements.

(1) Creation of the steering group – at the appropriate time

Depending upon its size and complexity the Turnaround may require a preparation period of anywhere from 3–18 months, and the steering group should be set up before the start of the preparation phase. Apart from the amount of planning and preparation required, this time scale is also determined by such things as the need to purchase expensive proprietary items which are on a long delivery (e.g. in some parts of the world the delivery time for a compressor rotor, say, may be as much as 16 months), the requirement for pre-shutdown work that has to be completed well in advance (e.g. the excavation and installation of foundations with a long cure-time) and the design and procurement of pre-fabrications. If the group do not have enough time to make their decisions the Turnaround may be adversely affected, sometimes catastrophically so.

(2) Appointment of a full time Turnaround manager

The complexity of the Turnaround process and the fact that the group members can only operate on a part-time basis (because they still have their normal jobs to do) make it imperative that the group appoint a full time Turnaround manager who will act as the steering group's agent in all matters. The group invest authority in him to act on their behalf and to make decisions – within the confines of the policy as laid down by them. Once appointed, he should take over as chairman of the group and set the agenda for each meeting.

The Turnaround manager becomes the 'active' member of the group and ensures that issues are resolved by the *Proposal–Discussion–Decision–Action–Feedback* routine. That is, he proposes a course of action; the group discuss it; they take a joint decision on the course of action; he ensures that the action is carried out and then reports the results of the action back to the group (a detailed description of the Turnaround manager's role is given later in this chapter).

(3) Creation of a formal agenda

In order for the steering group to address the myriad aspects of the Turnaround they require a formal agenda which covers all critical points. Typically, this would include (but not be limited to) the following:

1.   CONSTRAINTS
1.1 Review of previous actions (including review of previous events)
1.2 Window of opportunity and timing constraints
1.3 Duration
1.4 Budget and cost estimates (Are there differences? How can they be resolved?)
1.5 Comparison of workloads (Is this event similar to previous events – or not?)
1.6 Availability of resources
1.7 Special constraints

2   OBJECTIVES
     Quantity–Quality–Time–Money–Safety

3   WORK SCOPE
3.1 Legal requirements – if any
3.2 Major tasks
3.3 Preventive tasks
3.4 Corrective tasks
3.5 Condition monitoring
3.6 Modifications
3.7 Projects
3.8 Demolition

4   PREPARATION
4.1 Preparation plan
4.2 Work specification
4.3 Material procurement
4.4 Contractor selection
4.5 Integrated Turnaround plan
4.6 Turnaround organization
4.7 Site logistics
4.8 Cost profile
4.9 Safety

A formal agenda ensures business is carried out in a rational manner and reduces uncertainty.

(4) The holding of regular group meetings

The best way to gain commitment from group members and drive the progress of the Turnaround is to hold regular formal meetings, the frequency of which will depend upon the preparation time available and the size and complexity of the event. For a large event the group might meet weekly for the first four weeks to clear off the bulk of early work. Then monthly until one month before the event, when it would revert to weekly.

(5) The minuting of every action and decision

Every meeting must be formally minuted, special attention being given to two particular elements, as below.

(i)   When any action is delegated it should be recorded in an action log in full together with the name of the person who will carry it out and the required completion date. If the person responsible does not attend the group meetings, the Turnaround manager must undertake to inform that person, in writing, of the requirements of the group. At each meeting the first point on the agenda must be a review of the action log to ensure that each one has either been completed or is progressing satisfactorily; if not, the reasons must be discussed and decisions taken as to how best to resolve the impasse.

(ii) Once decisions are agreed upon they should be written down formally, and every member of the group should be committed to them even if the agreement is not unanimous. The time for argument is at the discussion stage, not after a decision has been made.

There is a favourite saying among auditors 'if it wasn't written down it didn't happen'.

(6) Formulation of a preparation plan (see Figures 2.3, 2.4(a) and 2.4(b))

Because the planning and preparation of the Turnaround are carried out over a relatively long period of time, and include a large number of critical activities, the group requires a preparation plan, to monitor progress on a regular basis. Figure 2.3 shows the key events in the initiation and preparation phases, including both *strategic* events, that are the responsibility of the senior management and steering group, and *tactical* events which are the responsibility of the Turnaround manager. Figure 2.4(a) is in the form of a Gantt chart where the timings and durations of Turnaround team activities are displayed. Figure 2.4(b) is a network plan display with activities on the lines and due dates written in the nodes. What the three diagrams have in common is that they all define what has to be done, when it has to be done by and the dependencies between activities. Over the period of preparation the steering group, and more especially, the Turnaround manager, will use these documents to monitor progress and take any necessary action.

(7) The reviewing of progress on all issues

With so many issues at stake there needs to be a constant review of every aspect of the Turnaround process. Every steering group meeting should involve a risk assessment process for ensuring that activities completed to date are consistent with the current strategy and will, as far as possible, ensure a successful event.

The Turnaround manager

The Turnaround manager is the key role in the organization. It is his responsibility to ensure all activities are carried out in the initiation, preparation, execution and termination of the event. Whether this role is full time or not will depend on such factors as the size of the event, the complexity of the work etc., but for large events it is recommended that the role is full time. Depending upon the company's situation he may be:

- *A company employee* – where there is a member of staff with the relevant knowledge and experience.
- *A consultant manager* – if the company does not have anyone with the requisite knowledge and experience (several consultancies provide this service).
- *A manager from the main contractor* – if a contractor is being used to perform the work they can normally provide a suitable candidate.

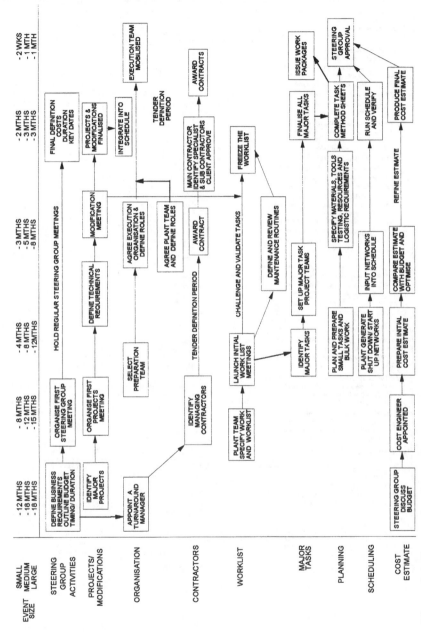

Figure 2.3 Preparation: key event programme

Project name					Start Date		Issue No................
							LEGEND Duration ▬▬▬
Compiled by		Signature			Date		Float 0000000
Approved by		Signature			Date		MUST date ▽ Dependency ....

Week 01 02 03 04 05 06 07 08 09 10 11 12 13 14 15 16 17 18 19 20 21 22 23 24 25 26 27 28

1 DEFINE WORKSCOPE & CONSTRAINTS
1.1 Collate work lists and projects.............
1.2 Validate & delete unnecessary work.............
1.3 Obtain approval and close worklist.............

2 PREPARE JOB SPECS (IF REQUIRED)
2.1 Select Turnaround planning team.............
2.2 Plan major tasks & bulkwork specifications.............
2.3 Issue for comment, amend & finalise.............

3 IDENTIFY PRE SHUTDOWN WORK
3.1 Order long delivery items & fabrications.............
3.2 Order pre shutdown services.............
3.3 Expedite, receive, store & protect 3.1 & 3.2.............

4 IDENTIFY & PROCURE MATERIAL & ITEMS
4.1 Define material & item requirements.............
4.2 Select suppliers & place orders.............
4.3 Expedite, receive, store & protect.............

5 DEFINE CONTRACT WORK PACKAGES
5.1 Split work into manageable packages.............
5.2 Define types of contractors.............
5.3 Create discrete contractor workscopes.............

6 SELECT TURNAROUND CONTRACTORS
6.1 Prepare & issue ITBs to contractors.............
6.2 Clarify & negotiate contracts.............
6.3 Select contractors & let contracts.............

7 CREATE INTEGRATED TURNAROUND PLAN
7.1 Receive & collate contractor work plans.............
7.2 Create, run, refine & re-run schedule.............
7.3 Issue for comments, amend and finalise.............

8 DEFINE TURNAROUND ORGANISATION
8.1 Define "black box" organisation chart.............
8.2 Create initial detailed organisation chart.............
8.3 Issue for comment, refine & finalise.............

9 CREATE SITE LOGISTICS PROGRAMME
9.1 Identify all site support requirements.............
9.2 Collate requirements into a rational plan.............
9.3 Order, receive & place requirements.............

10 FORMULATE TURNAROUND COST PROFILE
10.1 Formulate initial turnaround costs (±20%).............
10.2 Create detailed cost profile.............
10.3 Issue for comment, refine & finalise.............

11 ESTABLISH PERMIT TO WORK SYSTEM
11.1 Define permit to work philosophy.............
11.2 Select permit to work issue team.............
11.3 Block out permits pre shutdown.............

12 DEFINE SAFETY PROGRAMME
12.1 Define turnaround safety philosophy.............
12.2 Select team & define safe system of work.............
12.3 Set up safety organisation on site.............

13 DEFINE QUALITY PROGRAMME
13.3 Define turnaround quality philosophy.............
13.2 Select team & define methodology.............
13.3 Set up quality team on site.............

14 BRIEF TURNAROUND ORGANISATION
14.1 Collate individual briefing documents.............
14.2 Produce an integrated briefing package.............
14.3 Brief everyone in the organisation.............

Figure 2.4(a) Turnaround preparation programme

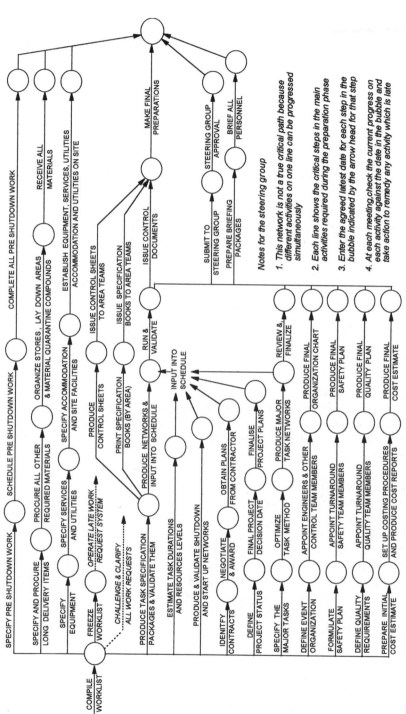

Figure 2.4(b) Turnaround preparation network

The following labels appear within the network diagram:

COMPILE WORKLIST

SPECIFY PRE SHUTDOWN WORK

SPECIFY AND PROCURE LONG DELIVERY ITEMS

SPECIFY EQUIPMENT

FREEZE WORKLIST

OPERATE LATE WORK REQUEST SYSTEM

CHALLENGE & CLARIFY ALL WORK REQUESTS

SCHEDULE PRE SHUTDOWN WORK

PROCURE ALL OTHER REQUIRED MATERIALS

SPECIFY SERVICES AND UTILITIES

PRODUCE TASK SPECIFICATION PACKAGES & VALIDATE THEM

ESTIMATE TASK DURATIONS AND RESOURCES LEVELS

PRODUCE & VALIDATE SHUTDOWN AND START UP NETWORKS

COMPLETE ALL PRE SHUTDOWN WORK

ORGANIZE STORES, LAY DOWN AREAS & MATERIAL QUARANTINE COMPOUNDS

SPECIFY ACCOMMODATION AND SITE FACILITIES

PRODUCE CONTROL SHEETS

PRODUCE NETWORKS & INPUT INTO SCHEDULE

INPUT INTO SCHEDULE

RECEIVE ALL MATERIALS

ESTABLISH EQUIPMENT, SERVICES, UTILITIES ACCOMMODATION AND UTILITIES ON SITE

ISSUE CONTROL SHEETS TO AREA TEAMS

PRINT SPECIFICATION BOOKS (BY AREA)

ISSUE SPECIFICATION BOOKS TO AREA TEAMS

RUN & VALIDATE

MAKE FINAL PREPARATIONS

ISSUE CONTROL DOCUMENTS

IDENTIFY CONTRACTS

NEGOTIATE & AWARD

OBTAIN PLANS FROM CONTRACTOR

DEFINE PROJECT STATUS

FINAL PROJECT DECISION DATE

FINALISE PROJECT PLANS

SPECIFY THE MAJOR TASKS

OPTIMIZE TASK METHOD

PRODUCE MAJOR TASK NETWORKS

REVIEW &, FINALIZE

DEFINE EVENT ORGANIZATION

APPOINT ENGINEERS & OTHER CONTROL TEAM MEMBERS

PRODUCE FINAL ORGANIZATION CHART

FORMULATE SAFETY PLAN

APPOINT TURNAROUND SAFETY TEAM MEMBERS

PRODUCE FINAL SAFETY PLAN

DEFINE QUALITY REQUIREMENTS

APPOINT TURNAROUND QUALITY TEAM MEMBERS

PRODUCE FINAL QUALITY PLAN

PREPARE INITIAL COST ESTIMATE

SET UP COSTING PROCEDURES AND PRODUCE COST REPORTS

PRODUCE FINAL COST ESTIMATE

SUBMIT TO STEERING GROUP

PREPARE BRIEFING PACKAGES

STEERING GROUP APPROVAL

BRIEF ALL PERSONNEL

Notes for the steering group

1. *This network is not a true critical path because different activities on one line can be progressed simultaneously*

2. *Each line shows the critical steps in the main activities required during the preparation phase*

3. *Enter the agreed latest date for each step in the bubble indicated by the arrow head for that step*

4. *At each meeting, check the current progress on each activity against the date in the bubble and take action to remedy any activity which is late*

Personality

If the endeavour is going to be successful the person selected for the role of Turnaround manager should possess the following character traits:

- Leadership, team building and negotiating skills;
- A good working knowledge of Turnaround type projects;
- Planning and co-ordinating ability;
- A flexible approach to complex problems;
- The ability to work effectively under pressure;
- A strongly developed sense of humour.

It takes patience and tenacity to manage the different individuals, systems and activities involved in the many difficult, and occasionally novel, situations that arise during the four phases of the Turnaround.

Roles and responsibilities

The specific roles and responsibilities of the Turnaround manager change as the project advances through its different phases. They include (but are not limited to):

INITIAL PHASE
- Attending the initial meeting of the working steering group;
- Advising the steering group on specific Turnaround requirements;
- Ensuring the collection and collation of basic data;
- Selecting and managing the Turnaround preparation team.

PREPARATION PHASE
- Convening all necessary meetings;
- Monitoring preparation, progress and information feedback to the steering group;
- Validating work scope and freezing the work list;
- Approving the selection of contractors and vendors;
- Organizing, optimizing, finalizing and presenting the following plans to the steering group for discussion, approval and action:
  - Turnaround schedule
  - Organization chart
  - Safety plan
  - Quality plan
  - Logistics plan
  - Cost estimate
  - Communication and briefing package
- Organizing the Turnaround briefing programme.

EXECUTION PHASE
- Organizing assistance for the plant team during plant shut down;
- Managing the daily conduct of the Turnaround;
- Taking action to ensure targets and objectives are met;
- Delegating specific responsibilities, tasks and activities;
- Chairing the daily control meetings;
- Reporting significant issues to the steering group;
- Setting priorities and trouble-shooting;
- Negotiating changes in resources and work scope;
- Organizing assistance for the plant team during plant start-up.

TERMINATION PHASE
- Demobilizing resources and equipment, and organizing the cleaning of the site;
- Co-chairing debrief sessions with the plant manager;
- Collating all information and writing the Turnaround final report.

ELAPSE PHASE
- This is the time that elapses between one Turnaround and the next, during which improvements can be made to ensure improved performance on the next event.

Throughout the changing phases of the lengthy event, the one constant key figure is the Turnaround manager.

The preparation team: membership (see Figure 2.5)

Purpose

During the initial phase of the Turnaround the manager appoints a preparation team who will be responsible to him for the detailed planning and preparation of all aspects of the project. It will consist of a small number of permanent members to carry out the general work, backed up by temporary members co-opted to perform specific functions. The size of the team will depend upon the size and complexity of the Turnaround. For present purposes it is assumed that we are dealing with a large event requiring a full strength team. For smaller events the team would be scaled down, with some of the roles combined for performance by one person, to suit the available funds and the lesser amount of work. In addition to the Turnaround manager, the permanent team would encompass those roles and responsibilities which will now be briefly summarized (but will be discussed in more detail later).

Note: The nomenclature used here is one which has been arbitrarily adopted by the author, there being no universally agreed one. It is the function which is important, not its name.

Roles and responsibilities

1. Turnaround manager
- selects the members of the preparation team
- ensures the team is properly resourced, equipped, safety inducted and introduced to plant personnel
- sets out the team's targets for preparation
- attends nominated preparation meetings
- resolves strategic issues for the prep engineer
- delegates authority to the prep engineer
- monitors team performance via the prep engineer
- reports directly to the turnaround steering group
- refers policy issues to the steering group and assists the team to resolve them

2. Preparation (prep) co-ordinator
- manages the daily conduct of the team
- chairs all preparation team meetings
- liaises with the client on a daily basis
- allocates work to team members
- validates the worklist and refers it to the turnaround manager for final approval
- plans all major tasks
- resolves operational issues
- monitors the team's performance and adjusts what is required to stay on target
- reports directly to the turnaround manager
- refers strategic issues to the turnaround manager

3. Planning co-ordinator
- assists preparation engineer to validate work list
- liaises with client planning organization
- allocates planning tasks
- supervises the work of the planners
- checks completed task sheets and forwards them to the plant manager for validation
- inputs networks and data into the scheduler
- creates and optimizes the turnaround plan
- assists the prep co-ordinator planning major tasks
- resolves all planning issues
- reports directly to the preparation co-ordinator
- refers operational issues to the prep co-ordinator

4. Planning team
- gathers basic data for planning from the plant
- plans and networks small tasks
- produces task sheets
- specifies bulkwork and lists on control sheets
- collates all documentation for task packages
- produces material requirement sheets
- produces equipment requirement sheets
- liaises with site logistics team
- prepares area task books for area co-ordinators and supervisors
- reports directly to the planning co-ordinator
- refers planning issues to the planning co-ordinator

5. Logistics co-ordinator
- liaises with plant personnel to gather basic data on the plant and available surrounding land
- draws up a master plot plan showing locations of all turnaround logistic requirements
- produces various versions of the site plot plan for use by different disciplines
- liaises with everyone involved in the turnaround
- arranges all outdoor lay down areas
- sets up turnaround stores and formulates receipt and issue procedures for materials and equipment
- procures/receives, locates, protects and maintains:
  - materials and proprietary items
  - tools and technical equipment
  - transportation and cranage
  - services and utilities
  - accommodation and facilities
- controls hazardous substances
- provides for the daily needs of all personnel
- allocates tasks to the logistics team
- supervises the daily work of the team
- monitors team performance
- reports directly to the preparation co-ordinator
- refers operational issues to the prep co-ordinator

6. Logistics team
Usually semiskilled personnel –
- reports to the logistics co-ordinator
- carries out all tasks set by the site logistics co-ordinator

Figure 2.5 The preparation team

Preparation co-ordinator

Whereas the Turnaround manager has many duties which require him to be mobile, the preparation co-ordinator would be permanently located on plant with the team. Typical roles and responsibilities would include, but not be limited to:

- Reporting directly to the Turnaround manager;
- Liaising with plant and other necessary personnel on a daily basis;
- Managing the daily conduct and performance of the team;
- Organizing and chairing all preparation team meetings;
- Allocating work packages to team members;
- Validating the work list and referring it to the Turnaround manager for approval;
- Planning all major tasks;
- Resolving day-to-day operational issues on behalf of the team;
- Monitoring team performance and adjusting requirements where necessary;
- Referring strategic issues to the Turnaround manager for resolution.

On small events the Turnaround manager would fulfil this role.

Planning co-ordinator

The planning co-ordinator works closely with the preparation co-ordinator. Typical roles and responsibilities would include, but not be limited to:

- Reporting directly to the preparation engineer (or Turnaround manager)
- Liaising with plant planners and other necessary personnel;
- Assisting the preparation co-ordinator with validation of the work list;
- Allocating planning tasks;
- Supervising the work of the planners;
- Checking completed planning documents and passing them to plant personnel for validation;
- Checking critical path networks produced by the planners and putting them into the schedule;
- Creating and optimizing the Turnaround schedule;
- Assisting the preparation co-ordinator to plan major tasks;
- Resolving all planning issues on behalf of the team;
- Referring operational issues to the preparation co-ordinator for resolution.

Planning team

The number of planners is dictated by the amount of work to be done and the time and money available. Typical roles and responsibilities would include, but not be limited to:

- Reporting directly to the planning co-ordinator;
- Gathering basic data for planning from plant and other necessary personnel;

- Producing task specifications and networks for all small tasks;
- Specifying bulkwork tasks and listing them on control sheets;
- Producing material requirement sheets for specified tasks;
- Producing equipment and services requirement sheets;
- Liaising with logistics team;
- Preparing task books for execution team superintendents and supervisors;
- Referring planning issues to the planning co-ordinator for resolution.

On small events there may be only one planner who fulfils both of the above roles.

Logistics co-ordinator

Site logistics is concerned with the reception, location, protection, distribution and final disposal of all items and services required for the execution of the Turnaround. Typically, the logistics co-ordinator roles and responsibilities would include, but not be limited to:

- Reporting directly to the Turnaround manager;
- Defining the plant boundaries and the available surrounding land;
- Drawing up a site plot plan showing all logistics requirements;
- Setting up stores, lay-down and quarantine areas;
- Receiving on site, locating, protecting, distributing and disposing of:
  - materials, consumables and proprietary items
  - tools and equipment
  - vehicles and cranage
  - services and utilities
  - accommodation and facilities
- Controlling hazardous substances;
- Providing for the daily needs of all personnel;
- Allocating work to the logistics team;
- Supervising the daily work of the team;
- Monitoring team performance;
- Resolving logistics issues;
- Referring operational issues to the preparation co-ordinator for resolution.

Logistics team

This team is normally made up of semiskilled and non-technical personnel such as storemen, drivers, cleaners etc. They report to the logistics co-ordinator and carry out the tasks set by him. Again, for small events, these activities may be performed by one person.

Preparation team: gathering basic data

Once the preparation team has been appointed they are normally accommodated in or near the plant offices because they will spend a great deal of their time on the plant. Manufacturing plants are hazardous places and it is likely that some or all of the team will be unfamiliar with the plant. If the term is not made up of permanent plant personnel it is the responsibility of the Turnaround manager to arrange the following safety precautions for any team members who are not existing members of the plant team.

Plant safety induction

The safety induction should familiarize the team with the hazards presented by the plant and with the operation of its emergency procedures (include the various sounds of the fire, toxic release and evacuation sirens). It must give adequate information regarding the permit-to-work system because members of the team may have to visit parts of the site controlled by permit in order to assess a particular job.

Plant tour

Experienced personnel should conduct the team on a tour of the plant, to familiarize the members with its geography and its hazardous areas.

Temporary site pass

This will last for the duration of the team's stay and will ensure that they are formally recorded as working on the plant.

Provision of safety gear

To ensure the team can be dressed to withstand the normal hazards of the plant they must be equipped both with normal safety gear (coveralls, hard hat, steel-toe-capped boots and light eye protection) and any plant-specific 'special' safety gear (which may include goggles, gloves, respirators etc.).

Data gathering

A Turnaround is a complex process and like any other the quality of its outputs depends upon the quality of its inputs. The inputs and outputs of the preparation team are information. Data gathered from plant personnel and records are transformed into the specifications and plans required to carry out the Turnaround. Unfortunately, on many plants, especially older ones where there have been several turnovers in personnel – and many modifications to equipment, procedures, systems and processes – some necessary data may either not exist or, if it does, may be out of date or unreliable in some other way. It is therefore essential that the basic data gathered from the plant and other sources is checked and validated by the team members before being used. There are two basic ways to gather data from people: by personal contact and through formal meetings.

Personal contacts (see Figure 2.6)

In order to gather accurate data, a network of personal contacts should be built up. The personnel involved would typically include – but not be limited to – Plant Manager, Safety Officer, Engineering Manager, Maintenance Manager, Project Manager, Chief Inspector, Stores Controller, Process Supervisors, Maintenance Supervisors, Electrical and Instrumentation Engineers and Supervisors. Rigging Supervisor, Scaffolding Supervisor, Civil Supervisors and the Plant Maintenance Planning Team Leader. These are the people who

Meetings to be kicked off include, but are not limited to:

		Comments
1.	Steering group meeting	To set policy and strategy. To monitor progress and performance.
2.	Plant standards meeting	To establish the standards for Turnaround work.
3.	General worklist meeting	To justify, clarify and categorize all work requests.
4.	Major task review meeting	To review the method of work for all major tasks.
5.	Project review meeting	To review progress an all projects.
6.	Inspection review meeting	To finalize details & requirements for all inspection tasks.
7.	Spares review meeting	To ensure all spares, especially long delivery items, are ordered and expedited.
8.	Shutdown/start-up meeting	To integrate the schedule for the Turnaround.
9.	Safety review meeting	To generate a safe system of work.
10.	Quality review meeting	To set quality standards for Turnaround work.
11.	Site logistics meeting	To acquire all of the data needed to plan the Turnaround.

Figure 2.6 Basic data gathering

possess the information the preparation team requires; building solid working relationships with them is important. Another issue to consider is that on any plant there will be a small number of individuals (sometimes just one person) who have worked on it for a long time and understand many of the ways that it works and why it sometimes doesn't. They will know the history of the plant and will be aware of its chronic problems. They will know where records are and will be able to advise the team on the validity of data. It is vital that they be sought out, recognized, cultivated and their knowledge and experience exploited to the benefit of the team.

Formal meetings (see Figure 2.7)

The very mention of the word 'meeting' tends to send a chill down the spine of most people. Meetings are often seen as wasteful, time consuming and ineffective. Turnaround preparation requires many formal meetings because they remain the best (and may be the only) way for a group of people to transmit, check, challenge and validate information.

There are a number of principles which guide and enhance the effectiveness of preparation meetings.

First – the agenda of each meeting should be kept as simple as possible so that attendees can focus their minds on particular requirements.
Second – only those who are directly involved with that particular aspect of preparation should be invited to attend.
Third – each meeting should have either a clearly defined block of work to accomplish, a defined time limit, or both.
Fourth – the meetings should all be chaired by the appropriate member of the Turnaround preparation team, who then sets the agenda and takes the minutes.

Some subjects (e.g. plant standards) may require only one meeting, while others, involving the resolution of complex issues (e.g. work listing) might require the convening of a number of meetings.

Typically, the meetings involved in Turnaround preparation would include, but not be limited to, the following subjects: Turnaround policy; plant standards; work list; major task review; project review; inspection review; spares review; safety review; quality review and site logistics. In addition to these there will be a large number of *ad hoc* meetings, convened to resolve specific issues (e.g. the resolution of single design, manufacturing or delivery problems).

Documentation (see Figure 2.8)

There may also be a large number of documents involved in preparing a Turnaround. These may be either in hard copy or electronic. They will include, but not be limited to, plant organization charts, piping and instrumentation diagrams, technical library, spares lists, previous Turnaround

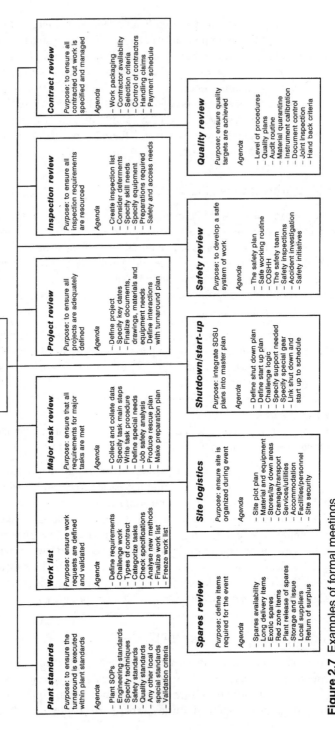

Formal meetings

Plant standards

Purpose: to ensure the turnaround is executed within plant standards

Agenda

– Plant SOPs
– Engineering standards
– Specify techniques
– Safety standards
– Quality standards
– Any other local or special standards
– Validation criteria

Work list

Purpose: ensure work requests are defined and validated

Agenda

– Define requirements
– Challenge work
– Types of contract
– Categorize tasks
– Check specifications
– Analyse new methods
– Finalize work list
– Freeze work list

Major task review

Purpose: ensure that all requirements for major tasks are met

Agenda

– Collect and collate data
– Specify task main steps
– Write task procedure
– Define special needs
– Job safety analysis
– Produce rescue plan
– Make preparation plan

Project review

Purpose: to ensure all projects are adequately defined

Agenda

– Define project
– Specify key dates
– Finalize documents, drawings, materials and equipment needs
– Define interactions with turnaround plan

Inspection review

Purpose: to ensure all inspection requirements are resourced

Agenda

– Create inspection list
– Consider deferments
– Specify skill needs
– Specify equipment
– Preparations required
– Safety and access needs

Contract review

Purpose: to ensure all contracted out work is specified and managed

Agenda

– Work packaging
– Contractor availability
– Selection criteria
– Control of contractors
– Handling claims
– Payment schedule

Spares review

Purpose: define items required for the event

Agenda

– Spares availability
– Long delivery items
– Exotic spares
– Red zone items
– Plant release of spares
– Storage and issue
– Local suppliers
– Return of surplus

Site logistics

Purpose: ensure site is organized during event

Agenda

– Site plot plan
– Material and equipment
– Stores/lay down areas
– Cranage/transport
– Services/utilities
– Accommodation
– Facilities/personnel
– Site security

Shutdown/start-up

Purpose: integrate SDSU plans into master plan

Agenda

– Define shut down plan
– Define start up plan
– Challenge logic
– Specify support needed
– Specify special gear
– Link shut down and start up to schedule

Safety review

Purpose: to develop a safe system of work

Agenda

– The safety plan
– Safe working routine
– COSHH
– The safety team
– Safety inspections
– Accident investigation
– Safety initiatives

Quality review

Purpose: ensure quality targets are achieved

Agenda

– Level of procedures
– Quality plans
– Audit routine
– Material quarantine
– Instrument calibration
– Document control
– Joint inspection
– Hand back criteria

Figure 2.7 Examples of formal meetings

Data required includes, but is not limited to:

Comments

Item	
1. Plant organization chart	
2. Plot plan of the plant	
3. Previous turnaround reports	
4. Previous budget/cost report	
5. Plant permit to work system	
6. Statutory inspection list	
7. Mechanical work list	
8. Electrical work list	
9. Instrument work list	
10. Trip and alarm schedule	
11. Civil work list	
12. Machine tasks	
13. Equipment cleaning tasks	
14. Projects and modifications	
15. Shutdown/start-up network	
16. Access to plant P and IDs	
17. Access to line diagrams	
18. Access to plant history	
19. Access to technical library	
20. Access to spares lists	

Figure 2.8 Documentation

technical and cost reports, permit to work procedures, statutory inspection list, mechanical/instrument/electrical/civil work lists, machines task list, equipment cleaning tasks, trip and alarm schedule, shutdown and start-up networks. The Turnaround planning team needs to gain access to this documentation.

Conclusion

The initiation phase is a period during which the groundwork is laid for the preparation phase. The preparation team is selected, re-located to the plant site if necessary, and inducted. The team members then make contact with key people on and off the plant, organize the initial meetings for all the various aspects of the Turnaround and begin to acquire the necessary documentation and other information which will allow them to plan and prepare for the event. It is, as it were, the foundation of the Turnaround process.

Case study

The total responsibility for the Turnaround on chemical plant A – one of a group of five owned by the same company – fell on the maintenance manager. The senior management's involvement was to allocate a budget based on what was spent on the last Turnaround, three years previously, plus a sum to cover for inflation and to appoint the maintenance manager as Turnaround manager. Thereafter he was left to get on with it.

The work list was formulated by plant personnel, and the job of the Turnaround planning team (who had been seconded from the other four plants in the group) was seen as simply planning and scheduling the work. There was little contact between the maintenance manager and the plant manager, both of whom had been in their jobs for less than two years. The Turnaround plan was prepared and a cost estimate generated one week before the start of the event. The estimate for the event was calculated at 64 per cent above the budget allocation and the planned duration was three days longer than the time allocated by the business manager for the event.

Three days before the event was due to start an emergency meeting was finally called. It was attended by the plant manager, the maintenance manager and the divisional director for production. The last of these demanded to know why the cost of the Turnaround had risen so sharply. The maintenance manager explained that there were several large packages of work which were only carried out every second Turnaround so they had not been included in the previous one (from which the budget had been formulated). He had only found this out when he had investigated the reason for the sharp increase in cost. No one had informed him of the significant difference in work scope between the previous and present events.

The director asked if the extra work could be eliminated from the event. The maintenance manager replied that he just did not know at this stage and there was not enough time to do a worthwhile assessment.

The director then asked the plant manager if he could contribute anything useful. He replied that he also could not say whether the work needed doing or not – in any case it was a maintenance issue not a production one. He did offer the advice that if someone in the past felt it was necessary to do it then it probably was needed, and if it was not done the consequences might be serious. The director resigned himself to allowing the work to be done and left the meeting after roundly castigating both managers for their incompetence – he was not looking forward to the meeting with the board at which he would have to inform them that there was going to be a serious budget and duration over-run. ... A month later, the story was worse. Because a large amount of emergent work had arisen the final bill was 97 per cent greater than budget and there was an over-run of five days.

In the light of what has been said in these first two chapters, the flaws in strategy may be glaring and obvious, but the questions which have to be addressed are:

- What did each of the three principals do (or not do) which contributed to the failure?
- What could they have done to prevent it?

The ultimate responsibility lay with the senior management of the company, but the blame was laid squarely at the feet of the maintenance manager. He was moved to another job.

3

Validating the work scope

Introduction

A Turnaround is a task-oriented event and the list of tasks – the *work scope* – is the foundation upon which all other aspects of the event rest, especially safety, quality, duration, cost, resource profile, material and equipment requirements.

Initially there is no work scope, only lists of work requests generated by the production, maintenance, engineering, projects and safety departments and the like. At this stage the work may be either well or badly defined and it may or may not be necessary. The job of the plant and Turnaround teams is to take these basic work lists and, using a technique known generally as *validation*, get the work requests properly defined and weed out the unnecessary work.

Unnecessary work

Why should there be unnecessary work? Surely the people generating the work lists know what needs to be done and would not complicate matters or generate extra costs by requesting work that was unnecessary?

To address this issue it is necessary to recall the principle, already referred to, of performing the absolute minimum amount of work required to protect plant reliability. A Turnaround is a maintenance event (in part) and so should conform to this principle. Work lists, however, are not generated by business managers – who have an eye on the bottom line, the profit margin – but by people who work on the plant (or are closely associated with it) and who form personal judgements about the state of the equipment, the reasons for its various malfunctions, its chronic problems and what should be done to rectify these faults.

These judgements may be based on objective evidence and measurement but they may also be subjective, based on no more than a gut feeling. In either case, the conclusions drawn about causes of problems – and especially about what has to be done to put things right – may be faulty and should be tested against objective criteria to ensure their validity and confirm the necessity for carrying out the work requested.

Another category of unnecessary work is that which could (and should) be done either while the plant is on line or during some other outage. A popular perception among plant personnel is that

'The plant is shut down anyway, so why not get as much work done as possible?'

Which seems at first glance to be a reasonable approach and is, in fact, one which is adopted in some companies. The reason given by both maintenance and production managers is that they do not have enough staff to carry out the tasks during normal operation of the plant. If this is so then it is a short-coming in company staffing policy that is being masked by the Turnaround. If work is being dumped into the Turnaround then it is not being performed when it should be. This makes a nonsense of the maintenance programme. It should be recognized by the steering group and, if it is acceptable to them, a separate budget should be created and visibly labelled (for all to see) as 'Cost of "Dumped" work'.

There are several ways in which unnecessary work impacts upon the Turnaround and it is important to understand them if such work is to be challenged. They fall under the following headings:

(i) Increased expense

Because a Turnaround requires a large control team, work done during the event is more expensive than that carried out at any other time of the year.

(ii) Unofficially transferred work

If work is transferred unofficially from the routine maintenance list to the Turnaround work list it decreases the expenditure on routine maintenance (which is good for the plant manager) and increases the cost of the Turnaround (which is not so good for the Turnaround manager). Plant personnel who make such unofficial transfers of work often forget to transfer the funds set aside for the work in the maintenance budget to the Turnaround budget. Even if they did it would not cover the cost, because of reason (i). They may even forget to inform the Turnaround team that the work could be performed at another time.

(iii) Overcomplicated work requests

Even if the work requested is necessary, the method requested may not be the most cost effective. For example, the request 'Remove 36″ valve from top of column and overhaul' – no mean undertaking – might be reduced, after analysis by the Turnaround team, to 'Re-pack valve in-situ', a far easier and much less costly alternative.

(iv) Duplication

Because work can be requested by a large number of plant and other personnel, another category of unnecessary work can arise, namely work from duplicated requests. Duplication arises when different people request the same job. On many plants there is a three-shift rotation system and on some this may involve up to seven shifts of personnel. Communication between the shifts and between shift and day workers may not always be effective.

Two requests for the same work can initially be mistaken for two different jobs because they have been written differently. Compare:

'Investigate fault on distillation column isolation valve and repair as necessary'

with the alternative request:

'Overhaul IV 3024'

If the work is not validated at the start, a lot of time and money may be wasted planning and resourcing unnecessary work.

(v) Redundant work

This is work that is requested but is made unnecessary by another work request which is not cross-referenced (see the case study at the end of this chapter). It can also involve work requested for items of equipment that are no longer on the plant.

(vi) Nice-to-do work

This is work for which there is no business justification, but someone, somewhere, is determined to get it done for their own comfort. As one plant supervisor said, 'OK, there was no indication that the cooling system was silting up but I thought – because the plant is shut down, anyway – it would be nice to take the end cover off the heat exchanger and have a quick look inside'.

The job was done and no problems were found. But a problem did arise when it came to boxing-up the exchanger. The fastening studs had been accidentally removed and scrapped and a new set had to be ordered. The supervisor got his 'nice to do' job done – at a cost of $3000!

(vii) Desperation work

This is the 'We have to do something!' category. Consider this: a large electric motor had run continuously for ten years without any trouble and there was no indication of any current problems. However, the electrical engineer was anxious that the motor would fail before the next Turnaround (scheduled for two years ahead, exactly because it had run trouble free for so long!).

The motor was shut down, stripped and certain parts replaced with new – even though they showed no indication of excessive wear or damage. The unfortunate ending to this tale is that the motor burned out when it was re-started because the wrong type of bearing had been used to replace the original one. The duration of the Turnaround was extended by two days and over half a million dollars' worth of production lost.

(viii) Blackmail work

This category normally occurs late on in the event, usually very close to start up when people are operating under severe time pressure and are on their last reserves of energy. It is usually presented by a member of the plant team who uses his local knowledge and the fear of duration over-run to force a task onto the work list.

A typical statement is 'If you don't do this, then . . .' followed by any number of dire conclusions, e.g. 'The plant won't start up', 'Production will be affected', 'Product quality will suffer'. All of these may or may not be true but the question has to be asked – if such dire consequences are attached to not doing the work – why was it omitted from the work list in the first place?

Unnecessary work wastes money and time, clogs up the schedule, diverts scarce resources from critical work and may introduce faults into an otherwise reliable plant. The worst case scenario is that the unnecessary job becomes the critical-path job (as in the case of the electric motor).

It is therefore vital that the raw work lists are subjected to validation, which should proceed via the following basic steps:

CHECK
- that the request has been approved by a manager or his nominee;
- that the work is not duplicated or redundant;
- that the data on the request are accurate and describe exactly what is being requested.

CHALLENGE
- the need for performing the task at all;
- the need for performing the task during the Turnaround;
- the need for performing the task as requested (can it be done more economically?).

ANALYSE
- the safety/quality/material/equipment/resource requirements.

CLARIFY
- agree on the final wording of the work request;
- record the work request on the approved work list.

The validation process is employed to ensure, as far as possible, that the approved work scope contains only what is necessary to restore, maintain or enhance the reliability of the plant and which cannot be done at any other time (see Figure 3.1, a validation routine used by an oil refinery).

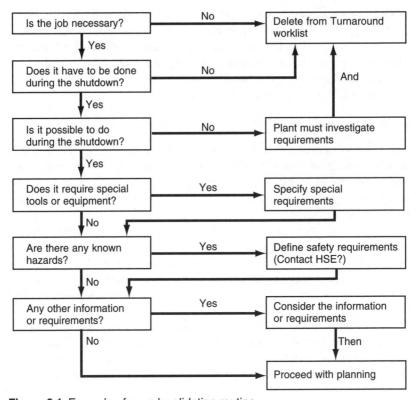

Figure 3.1 Example of a work validation routine

An effective work scope is achieved by analysing every task, which can be done at the following series of meetings.

The work list meeting

The effectiveness of the work list meeting will set the pace for all other activities on the Turnaround. Speed, as well as accuracy, is important.

Purpose

The meeting gathers together, and validates, the work requests generated by a large number of people in various departments, creating an approved work list for the Turnaround.

Participants

Should include:

- Turnaround manager (who is also the chairman);
- Preparation co-ordinator (if one is being employed);
- Planning co-ordinator and relevant planners;
- Plant or production manager;
- Maintenance manager (or whoever is responsible for the maintenance function);
- Plant supervisors;
- Electrical/Instrumentation engineers and supervisors;
- Nominees of either plant or Turnaround manager.

Work scope elements (see Figure 3.2)

A typical work scope would comprise the following categories of work:

- Statutory or company inspection requirements;
- Preventive maintenance;
- Corrective maintenance on known defects;
- Plant cleaning routines;
- Safety and quality initiatives;
- Work generated by the Turnaround.

If there is a very large or complex work list the decision may be taken to split it into sections and conduct several meetings rather than a single general one, the work being divided by geographical area, functional unit or craft discipline. If this is done there must be at least one final meeting to draw all the sub-lists together into a single list for final validation.

Remit (see Figure 3.3)

The remit of the meeting is as follows:

- Meet weekly (or to another agreed schedule);
- Gather together every job request;
- Eliminate unnecessary work;
- Justify tasks which stay on the work list;
- Clarify requirements for each task;
- Categorize all tasks;
- Generate actions to drive the process;
- Assign the action to appropriate personnel;
- Formally minute all actions, decisions, progress and problems.

Check list of issues requiring consideration

Statutory/company requirements
- Inspect/repair pressure vessels
- Inspect/repair registered pipe work
- Inspect/repair relief rtreams/readers
- Overhaul/replace relief valves
- Inspect/replace bursting disks
- Inspect/repair of ultimate protection devices
- Inspect/overhaul non-return valves
- Inspect/replace bursting disks
- Inspect/repair suspect pipe work
- Pressure test to expose suspected leaks
- Inspect pipe work for suspected fouling

Repairs must be specified by the turnaround technical team and approved by the turnaround management team

Preventive maintenance
- Overhaul rotating machinery
- Overhaul electric motors
- Check equipment alignments and adjust
- Check instrument calibration and adjust
- Overhaul/replace control valves
- Inspect flameproof equipment and repair/replace
- Check vessel linings and internal equipment and if necessary repair
- Investigate suspected malfunctions
- Lubricate/change oil/top up oil reservoirs

Repairs must be specified by the turnaround technical team and approved by the turnaround management team

Corrective maintenance – known defects
- Repair tube or pipe leaks
- Repair flange and joint leaks
- Repair/replace steam traps/lutes/pots
- Repair furmanited joints
- Repair other temporary fixes
- Repair/replace - instruments - pumps
  - motors - gearboxes - valves - cable runs
- Repair/replace - handrails - access ladders
  - platforms - insulation
- Large asset repair
  - Compressors/turbines/large pumps
  - Vessel shells - internals - externals
  - Column trays and internals
  - Fin fan coolers - Cooling towers
  - Heat exchangers

Plant cleaning
- Cleaning requirements
  - Heat exchangers - vessels - pipelines
  - Equipment removed for overhaul
  - Cooling tower - Fin fan coolers
  - Drains - reservoirs - sumps - pits
- Cleaning techniques
  - Water washing - Jet washing
  - Chemical cleaning
  - Grit blasting/Slurry blasting
  - Manual cleaning
  - Mechanical cleaning
- Cleaning locations
  - In situ
  - On site water washing bays
  - On site decontamination bays
  - Off site cleaning facilities

Safety and quality initiatives
- Removing asbestos insulation
- Inspecting pipe supports
- Removing redundant plant
- Re-routing pipe, work
- Special safety checks on equipment
- Complying with new safety regulations
- Upgrading equipment (modernizing)
- Planned gasket upgrading
- Upgrading foundations/roads etc.

Work generated by the turnaround
- Process washing and sweetening
- Isolating/de-isolating
- Safety checks prior to vessel entries
- Scaffolding - removing insulation - re-insulating
- Disposing of redundant plant and equipment
- Disposing of waste and hazardous substances
- Loop checks/trip and alarm checks
- Site cleaning and housekeeping
- Peripheral tasks
  - Interfacing with other production units
  - Working on line (stoppling etc.)
  - Battery isolations
  - Controlling shared services

Figure 3.2 Work scope elements

The major task review meeting

Major tasks need to be properly planned and prepared, because at least one of them will define the duration of the Turnaround and any of them, if not properly planned and prepared, may go wrong and become the critical path.

Check list of issues requiring consideration

Gather together every job request
- Ensure all work lists have been submitted
  - Statutory inspection – Company inspection
  - Major tasks – Plant modifications
  - Project work – Machine overhauls
  - Preventive tasks – Corrective tasks
  - Equipment cleaning – Catalyst/packing
  - Instrument tasks – Electrical tasks
  - Safety initiatives – Quality initiatives
  - Peripheral work – Plant investigations

- Freeze the work list at the specified date

Eliminate unnecessary work
- Challenge client's practices
- Challenge every job - does it need to be done - or done this way - or done so frequently?
- Ask - What are the consequences of not doing this job? If the answer is 'none' eliminate it
- Defer work to smaller outages
- Eliminate extremely hazardous work or change the requirement to make it safer
- Eliminate unauthorized or duplicate work
- Eliminate 'nice to do' work unless the plant manager authorizes and accepts the costs

Justify jobs which stay on the list
- Ensure the job is technically feasible
- Ensure job is economically feasible
- Check due dates for all statutory work
- Ensure the job can be done in the turnaround
- If the job is a modification, ensure the requisite senior management approval is obtained
- If the job means working on live plant, ensure senior management approval is obtained
- If a job is likely to take longer than the fixed duration of the turnaround, inform policy team

Categorize work
- Categorize as follows
  - Major tasks – to be worked up by major task review team and planned by preparations engineer
  - Small tasks – to be planned by planning team and be given a separate job method sheet
  - Bulk work – to be planned by planning team and be packaged on schedule lists
- Make out three separate work lists

Clarify requirements for each job
- Is there any information to indicate the job has abnormal technical or safety features?
- Are there any special techniques that have to be employed to do the job?
- Do the plant management require the job to be done in a particular way due to local needs?
- Is there any evidence to suggest the job has been difficult to do in the past or needed any special techniques or equipment?

Generate actions to drive the process
- Don't leave anything to chance or guesswork
- A great deal of information not on the work orders will be needed to specify the task
- Clarify the needs into specific actions in order to progress the job, such as:
  - Request information from a known source
  - Find out where information can be obtained
  - Reorganize information into required format
  - Investigate or measure up on-site
  - Formulate a plan or strategy

Assign actions to appropriate person
- Ensure the person who is assigned the action is capable of carrying it out
- Ensure the action is specific
- Agree an end date with the assignee
- If the assignee requires any special resources to carry out the action, provide them
- At each meeting, review outstanding actions and if they are not completed, find out why!
- Take all reasonable measures to ensure actions are completed on time

Other considerations

Figure 3.3 Work list meeting remit

Purpose

The purpose of the major task review meeting is to ensure that large, complex or hazardous tasks are given the due consideration that their importance merits, so that a relevant specification for the task may be produced.

Participants

Should include:

- Preparation co-ordinator (the chairman);
- Plant or production manager;

- Maintenance manager;
- Safety representative;
- Other nominated plant personnel;
- Anyone co-opted for their specialist technical knowledge.

Remit (see Figure 3.4)

- Meet at agreed intervals until all issues are resolved;
- Obtain all information and documents necessary to plan the task;
- Describe the plant or equipment in detail;
- Define the task, step by step;
- Identify and resolve technical challenges;
- Carry out a task hazard study;
- Formulate a safety routine and a rescue plan;
- Define the following:
  - A preparation plan;
  - Inspection requirements;
  - Material, equipment, services and utility requirements;
  - Manpower requirements;
- Minute all actions, decisions, progress and problems.

The inspection review meeting

Because, in some countries, various inspections are required by law, it is vital that the tasks be specified in such a way as to achieve the inspectorate's objectives and that the specifications are auditable (because the inspectorate may wish to examine them).

Purpose

The inspection review meeting is held to ensure that all necessary statutory and internally generated inspection requirements are identified and defined – and met via the employment of the most relevant techniques.

Participants

Should include:

- Preparation co-ordinator;
- Planning co-ordinator and relevant planners;
- Chief inspector (optional);
- Inspectors assigned to the overhaul;
- Any other nominee.

Check list of contingencies to be considered

Information and documents

- Work order number
- Plant history of the equipment
- Line diagrams and P and IDs
- Inspection, test and IMI reports
- Operating information
- Process information
- Assembly and detail drawings
- Past turnaround reports
- Accident/incident reports
- Any other useful information etc.

Plant details

- Plant name and number
- Type of plant
- Fabrication materials
- Process and/or services fluids
- Temperature/pressure/flow
- Access or egress hazards
- Entry problems/requirements
- Residual substances
- Any known hazards
- Any other useful information

Task to be done

- State reason for task
- Produce a step by step plan
  - Main task stages
  - Initial logic network
  - Activities for each step
  - Duration of each step
  - Timescale for complete task
- Define if task is critical path
- Produce an overall plan

Technical challenges

- Erosion/corrosion/damage/ distortion/misalignment
- Design or project input
- Special material (long delivery)
- Special welding or machining
- Special skills or knowledge
- Specialist sub-contractors
- Special analysis techniques

Task hazard study

- Identify potential hazards at each stage of the task
- Define the potential loss which is associated with hazards
- If possible, eliminate hazard
- If elimination is not possible, take steps to guard against hazards

The safety routine

- Define special requirements:
  - Inert entry procedure
  - Working in confined spaces
  - Working at height/multi-level
  - Hot work/work on live plant
  - Multi level working
  - Need for entry guardians
  - Communication needs
  - Standby emergency services
  - Safety equipment

The rescue plan

In case of an accident on a work site with difficult access/egress
- Ensure there is a rescue team
- Alert the rescue team before work starts
- Ensure there's an escape route
- Ensure an inert body can be removed from the job site
- Ensure standby emergency services

Preparation plan

- Shut down/start up programme with durations specified
- Cleaning - washing - cooling
- Isolation requirements
- Ventilation required
- Removal of insulation
- Safety monitor and test
- Access and platforms
- Permits to work

Inspection needs

- Previous inspection reports
- Company/external inspectors
- Inspection technique (special)
- Timescale for inspection
- Platform – moveable/modified
- Lighting/sheeting out
- Calibrated instruments
- Radiography/other hazards
- Statutory requirements

Materials

- Specification
- Traceability
- Availability
- Long delivery
- Consumables
- Delivery/protection/storage/ lay down areas/issue
- Special needs (welding etc.)

Equipment

- Cranes/special handling gear
- Generators/compressors
- Wagons/low loaders/bogeys
- Disposal gear (catalyst etc.)
- Scaffolding, ladders etc.
- Special tools (torque gear etc.)
- Pressure testing equipment
- Delivery/lay down areas
- Servicing (if required)

Services/utilities

- Electricity/gas/water/fuel
- Water wash/grit blasting/ chemical cleaning
- Lagging/insulation/painting
- Cleaning/substance disposal

Manpower

- Number of people required
- Disciplines/trade split
- Training/briefing needs
- Experience needs
- Special skills or knowledge
- Company/external manpower
- Working hours/shift patterns
- Relief workers/stand-ins
- Special payments
- Industrial relations issues

Other requirements

Other requirements

Figure 3.4 Major task review meetings

Remit (see Figure 3.5)

- Meet regularly until all issues are resolved;
- Create a categorized inspection list;
- Explore the possibilities of deferring inspection (if desirable);
- List all inspectors' requirements for each job;
- Define inspection techniques to be used;
- Define types of inspectors required;
- Minute all actions, decisions, progress and problems.

Check list of issues requiring consideration

Categorized inspection list
Should be categorized as follows:

- Pressure vessels
- Relief streams
- Registered pipe work
- Nominated pipe work #
- Registered non-return valves
- Ultimate protection devices
- Any other nominated inspection #

 # Plant manager nominates

Deferred inspections
In certain circumstances, a case can be made for deferring inspection, if:

- Past inspection indicates no appreciable deterioration in condition
- Evidence from other plants in the same industry sector indicates no appreciable deterioration
- On-line monitoring indicates that there is no appreciable deterioration
- New on-line techniques make it no longer necessary to open up vessels etc.
- Plant performance or physical data shows that the element has not deteriorated

This option should be considered

Inspector's requirements
In order to carry out the job the inspector will need some combination of the following:

- Vessel atmospheric clearance certificate
- Permit to work
- Guardians/communications
- Assistants
- Protective clothing
- Safety gear
- Breathing apparatus
- Working platforms/access ladders
- Lighting/electric power
- Ventilation/fans/air movers
- Briefing on any special circumstances

Inspection techniques
Various, including, but not limited to:

- Visual inspection
- Dimensional measurement
- Dye penetrant
- Magnetic particle
- Ultrasound
- X-ray or Gamma radiography
- Bubble testing
- Thermal Imaging
- Photogrammetry
- Other specialized methods

Types of inspectors
Due to the various types of work done, a number of different types of inspectors may be needed, including but not limited to:

- Pressure vessel inspectors
- Welding inspectors
- Radiographers
- Registered pipe work inspectors
- Machine inspectors
- Specialist technique inspectors

Other considerations

Figure 3.5 Inspection review meeting

Project work review meeting

Although project work is performed on the same plant at the same time as the Turnaround, it may be controlled by several different departments. It is therefore vital that it is properly integrated into an overall schedule, identifying interactions with Turnaround work and avoiding conflicts.

Purpose

The meeting should ensure that project work, normally planned by the project department, is properly integrated into the Turnaround schedule so that all conflicts are resolved and requirements met.

Participants

Should include:

- Turnaround manager;
- Planning co-ordinator;
- Plant manager;
- Project manager (optional);
- Project engineers;
- Any other nominees.

Remit (see Figure 3.6)

- Meet regularly until project issues are resolved;
- Specify major and minor projects;
- Clarify key dates;
- Review documentation;
- Review material and equipment procurement;
- Define interactions with Turnaround work;
- Minute all actions, decisions, progress and problems.

The validated work scope

All other aspects of Turnaround planning, i.e. safety, quality, costs, materials, equipment and resource requirements are derived from the work scope. It is not possible to plan against an open-ended work list. Therefore, once the scope has been validated the work list is frozen on a date agreed by the steering group. Thereafter, any requests for work are authorized by the highest executive authority available and, if it can be justified and is approved, should be placed on a 'late work' list. The money, time and resources required to carry it out are extra to budget and are identified as such. Typically, there will

Check list of issues requiring consideration

Types of major projects

Various types of major project include, but are not limited to:

- Plant extension
- Major asset renewal
- De-bottlenecking
- Major plant modifications
- Major safety modifications
- Process Improvement initiatives
- Large construction activities
- Large civil activities

Value typically more then 150,000 US dollars

Types of minor project

Various types of minor project include, but are not limited to:

- Minor asset renewal
- Fixed cost reduction
- Safety initiatives
- Quality initiatives
- Packaged work
- Small design packages

Value typically less than 150,000 US dollars

Key dates for major projects

To ensure project is ready in time the following key dates should be conformed to:

Key project activity	Time before turnaround
- Initial project definition	18–24 months
- Initial project meeting	12–15 months
- Meeting with turnaround team	6–9 months
- Contract awarded	6 months
- Organize pre-shutdown work	6 months
- Project closure date	2 months
- Project plan issued	2 months
- Delivery of all material and items	1 month
- Briefing of project work force	1 week

Project documentation

To ensure everyone concerned understands the project requirement, review the following:

- Feasibility/operability HAZOP study
- Modification approval document
- Design and detail drawings
- Line diagrams
- Break-in drawings
- Materials lists and delivery schedules
- Equipment lists and delivery schedules
- Contractor documents
- Pre-shutdown work list
- Isolation register
- Project schedule for integration into turnaround plan

Review material and equipment procurement

It is essential that the correct material/equipment (M and E) is on site when it is needed, so ask:

- Has all long delivery M and E been ordered?
- Who is responsible for expediting M and E?
- Who will take delivery of M and E?
- What are the M and E storage requirements?
- Who will inspect M and E to ensure conformance?
- Will any M and E arrive after plant shutdown?
- Does any M and E require special handling?
- Is any M and E hazardous?
- Will contractors bring their own M and E?

Interactions with turnaround work

- Who will manage the project?
- Will the project be the critical path?
- Can the project be packaged with any other turnaround work?
- Is there any conflict between the Project and any other turnaround work?
- Will the project require the involvement of anyone already dedicated to the turnaround?
- Will contractors bring their own management, inspection and supervision?
- Will any specialist vendor require assistance from any turnaround personnel?
- Who will be responsible for removal of the redundant M and E generated by the turnaround?

Figure 3.6 Project work review meeting

be a steady flow of late work requests throughout both the preparation and execution stages of the Turnaround (those arising during the latter stage are termed 'emergent work', which is discussed in a later chapter). A validated work scope simplifies planning and preparation; late work complicates them.

Case study

A Turnaround work list for a chemical plant contained a request for the replacement of two double-block-and-bleed valves on an eight-inch line, the existing valves being unserviceable. The work was validated and planned, and new valves bought – at some considerable cost. On Day 4 of the Turnaround the two unserviceable valves were removed and replaced by two new valves. The job took two days to complete. On Day 6 (just as the previous job was nearing completion) the eight-inch line was cut out and replaced by a ten-inch line as part of an upgrading project – complete with two ten-inch double-block-and-bleed valves, brand new.

An investigation revealed that, although the project programme had been integrated into the overall event schedule from a timing, duration and resource point of view, the technical content of the job had not been discussed at any Turnaround meeting. The operations supervisor who had been involved in the definition of the up-grading project, and therefore knew that the eight-inch line would be removed, had not passed this information on to the senior operator who had requested that the eight-inch valves be replaced. The question is – what safeguards can be put in place to prevent occurrences like this?

The ultimate responsibility lay with the Turnaround manager. If he had ensured that all projects were technically defined at a meeting which was attended by some of the same people who attended the work list meetings, the duplicated work would have been identified.

4

Pre-shutdown work

Introduction (see Figure 4.1)

Although the main business of the preparation phase is to plan the work which will be performed *during* the event, part of the preparation effort must be dedicated to ensuring that all work which has to be completed *beforehand* (if the event is to be performed effectively) is also planned and executed. It must be identified, progressed and completed as indicated in the upper section of the preparation network shown in Chapter 1 (see Figure 1.2).

Long delivery items

A significant part of the planning and preparation of a Turnaround is the timely procurement of hundreds, or even thousands, of items – materials, spares, proprietary plant equipment and so on. Some of these – e.g. very large components of critical machines or parts made from exotic materials – will be on long delivery (as mentioned before, a delivery time for a compressor rotor might well be as much as sixteen months). In order to identify these, it is necessary to analyse the work list as early as possible to ensure sufficient time is allowed for the various procurements. Remember, that although the delivery time quoted is from receipt of order, the order will be transmitted through the company's procurement system – which has its own time scale that can sometimes run into weeks, or even months. Also, after delivery on site there will normally be a requirement for goods inward inspection – and if the item fails this it will have to be replaced!

Figure 4.1 Pre-shutdown work

What to look for

When analysing the work list, categories which should flag up the need to check on delivery time at an early date include, but are not limited to:

- Rotating equipment (or spares);
- Large or specialized valves;
- Non-standard equipment;
- Large quantities of non ex-stock items (e.g. reformer tubes or column trays);
- Exotic materials;
- Catalyst and packing;
- Items procured from foreign countries;
- Items needing special permits or certificates;
- Items requiring further work after delivery (assembling, painting etc.).

What to do

When long delivery items are identified the following procedure should be followed:

1. Make out a list of the items.
2. Find out if they have already been ordered.
3. If not, identify who will procure them and then instruct him to:
   - (i) check whether promised delivery is before the Turnaround (if it isn't to inform you of the date of delivery during the Turnaround);
   - (ii) expedite and regularly report on progress and problems;
   - (iii) (if practical) witness any vendor's tests and inspect item before it is delivered.
4. Arrange for storage or lay-down area (which may be on site or remote) and ensure the item will then be protected and maintained.
5. Arrange for any goods inward inspection and quick return of any damaged item to the supplying vendor.
6. If item will not be delivered until after the start of the Turnaround, calculate when it will be available for installation.
7. Report any major problems to the policy team (e.g. delivery times which will mean that installation of the items concerned will extend the duration of the Turnaround).

It will be up to the steering group to decide whether the work should be eliminated from the Turnaround list or modified, or whether the criticality of the task would justify the extension of the duration. However, even after that decision has been taken, one should either work to bring forward the delivery date, modify the schedule to accommodate the delivery time or specify a different task which will delay the necessity for installing the item (e.g. if the replacement is going to be late, can a damaged item be refurbished to last a few more months until the replacement can be installed?).

Prefabricated work

This is usually pipework or structural steel and can cause problems similar to those posed by long delivery items. In this case the materials or items which make up the fabrication may be on short delivery or even ex-stock. What takes up the time is the fabrication itself. The prefabrication may be done on-site by local or contract manpower or it may be done off-site by a company contracted to do the work.

Prefabrication may have no time pressure on it whatsoever and that in itself can be a problem. Due to the belief that there is plenty of time, it is constantly put off until it *does* become a problem. Therefore, the Turnaround manager should ensure that all prefabrication is carried out in a timely manner.

What to look for

- Large amounts of small fabrications (steam traps etc.);
- Large single fabrications;
- Fabrications of exotic materials;
- Fabrications with complex configurations;
- Fabrication which will have to withstand high pressure or load.

A list should then be drawn up of all special requirements, such as:

- Non destructive testing;
- Pressure testing;
- Linings, coatings or cladding.

What to do

1. Obtain a copy of the prefabrication documentation, i.e. isometric or other drawings, welding procedures, pressure testing procedures etc.
2. Check whether anything on the prefabrication bill of materials is on long delivery (if so, use the procedure described in Chapter 5).
3. Double check the fit. Get the planner to go on site and measure up the job to see if the configuration of the prefabrication on the drawing will actually fit into the available space. Designers often work from site drawings which may not show what is actually there.
4. Check all testing and inspection requirements and ensure the equipment needed will be on hand and properly calibrated (with accompanying certificates). The equipment itself may have to be procured and it may be on long delivery!
5. Check on the materials and items for fabrication, to identify those which require material or conformance certificates.

6. Check (before delivery) that the certificates have been requested from the vendor and (after delivery) that they have been provided and are correct.
7. Check (after delivery) that the configuration is correct.

As with long delivery, major problems must be reported to the policy team so that they can be resolved.

Specialist technologies

As the technologies of maintenance become ever more sophisticated they increasingly involve expensive equipment needing more specialization in their application. A large number of specialist companies have emerged to perform the necessary work. Although the rates they charge are high, the various applications of their technologies can save much time and money and are consequently worth considering. The following list is representative of these, but is by no means exhaustive:

*   Photogrammetry;
*   Thermal imaging;
*   Remote cameras, videos, introscopy;
*   Inert entries;
*   Bolt tensioning and torquing;
*   Laser alignment and measurement;
*   Sound wave signature measurement;
*   On site heat treatment;
*   Arc air/plasma/water jet cutting;
*   Metal spraying;
*   Application of ceramic metals;
*   Reformer servicing;
*   Vibro-testing;
*   Mobile gasket workshops;
*   Mobile water and jet washing facilities.

These are often provided by small companies or small divisions of large companies. There may be no relevant specialists in the particular country where the Turnaround is taking place, so they may have to be brought in from elsewhere with all the attendant difficulties of doing so. Because there are not many companies in any one field they are normally booked up many months in advance and do not have the flexibility to suddenly change either their manning levels or procedures to meet a particular demand. It is therefore important that they are contacted as early as possible and given accurate specification of the work required. If possible, they should be asked to come to the site and demonstrate their technology so that it may be judged whether they are appropriate to the need. Alternatively, another site where the technology is being applied could be visited to inspect it in action.

Once the decision to use the technology has been made, the routine (described in Chapter 5) for handling contractor packages should be followed.

Vendors' representatives

There are certain plant items – such as boiler packages or digital control systems – which are normally serviced by vendors' representatives. This is frequently linked to warranty and in some cases there is no alternative but to use them – and they will normally have a heavy workload. In order to ensure that they will be on hand when required (i.e. during the Turnaround), and can provide a good service, the following steps should be taken as early as possible during the preparation phase:

1. Check if they will be available for the Turnaround.
2. If not, ask them to recommend someone else.
3. Ask them if they are familiar with the plant concerned.
4. Inform them of any current plant faults.
5. Check that they have the necessary materials.
6. Find out what support they will require.
7. Acquire a work scope and duration from them.
8. If it is feasible, ask them to visit the site before hand to check the equipment before the Turnaround starts.

Services, utilities, accommodation and facilities

Although the management of these requirements will be discussed in detail in Chapter 8, it is worth stating here that all services (water washing, scaffolding, lagging etc.), utilities (water, electricity and gas), accommodation (offices, rest rooms, stores etc.) and facilities (canteens, washrooms, toilets etc.) need to be in position on site and serviceable before the Turnaround starts – so the planning, resourcing and execution of this undoubtedly qualifies as pre-shutdown work.

Case study: Red Zone

The management addressed the problem of long delivery items by creating a very simple but effective material 'Red Zone' routine to handle the many thousands of items required for the Turnaround – many of which were procured from abroad.

Firstly, the 'Red Zone Philosophy' was established. This stated that any item which was promised for delivery within four weeks of the start of the event (some items were not due to be delivered until after the start date) would be declared 'Red Zone'. These had to be handled by an appointed expediter, whose instructions were as follows:

1. Contact all suppliers or manufacturers of items in the Red Zone and, if possible, negotiate a new delivery time which will remove the item from the zone.
2. Agree, with each manufacturer, a progress and delivery programme with key dates which can be monitored.
3. Ensure the manufacturer understands the paperwork which will be required to accompany the item (test certificates etc. – these were listed).
4. Regularly monitor the progress of each Red Zone item (in some cases this may require a visit to the vendor's premises) to check that key dates are being met and that the promised delivery date will therefore be achieved.
5. Ensure that, as far as is practically possible, all required inspections or tests are carried out at the manufacturer's premises and, if necessary, appoint an independent inspectorate to represent the company (especially in foreign countries).
6. Alert the relevant engineer at the earliest date of any issue which is likely to cause a delay in the delivery of the item.
7. Check, just prior to delivery, that all paperwork has been generated and will accompany the item (or, if sent under separate post, will arrive before the item).
8. Negotiate (or appoint an agent to do so on behalf of the company) with customs officials in both the exporting and the importing countries to ensure that all exporting and importing regulations are complied with, thus ensuring that this issue will not delay delivery.
9. Check the entire delivery route to identify any potential hold ups.
10. At the earliest date, identify and calculate costs of alternative delivery routes for any item facing a potential delay – this to include an emergency system for items delayed at the last minute.
11. Sign off the item as delivered only when all goods inward checks have been completed and the resulting documentation is in hand.
12. Record all issues which caused – or which might cause – delays and make recommendations for expediting future programmes.

None of this is rocket science! It is merely meticulous attention to detail – attention which leads to the overall success of procurement.

At the end of the event the expediter's final report drew attention to several cases where, if the Red Zone had not been in operation, items would have arrived late, and two instances – both of which concerned large critical items – where the item would not have arrived at all. Oddly enough, the report was never published, so although the lessons were learned on an individual basis, they were not added to the sum of corporate knowledge.

5

Contractor packages

Introduction

It is not within the scope of this book to discuss the work of a company's procurement or contract departments. Most organizations have a well-developed strategy for organizing the contractor requirements for any project. However, the unique nature of a Turnaround project demands additional action by the Turnaround team. This requires, in turn, that the team has a basic level of commercial awareness and a working knowledge of the main issues surrounding the use of contractors.

In the context of Turnarounds 'contractors' are companies that specialize in performing some or all of the activities involved in an overhaul. They may be very specialized (e.g. an electronics company servicing a digital control system), multi-skilled (able to carry out mechanical, fabrication and civil work), specialize in the total planning and execution of events, or perform mundane tasks such as cleaning.

Contractors are all commercially aware companies in business to make a profit and, because of greater familiarity with Turnaround contracting, are normally more skilled at negotiation than the client company. The majority are reputable companies who make a valuable contribution to the business of Turnarounds and will perform work effectively so long as all the necessary preparations have been made by the client.

In many cases where a client company feels it has been 'ripped off' by a contractor the truth may be that the extra money it had to pay was the penalty for its own bad work specification, disorganization during the event and lack of the basic commercial awareness that, contractually, some things are possible and some are not. The contractor will honour the contract but he will also demand payment for all work he is asked to do – planned and unplanned.

Using contractors – the upside

There are several reasons why contractors are used, and the following list gives some of the main ones.

Resource availability

Turnarounds typically employ more people, and demand more skills, than are employed by the company to operate a plant.

Experience

Even where manpower is available on the plant, the fact that it may be only shut down once every two, three or even five years means that personnel do not get the opportunities to practise the skills required for executing Turnarounds and, as the Americans say, 'if you don't use it you lose it!' Contractors, on the other hand, are doing the same work week in and week out and should be more skilled at performing the required tasks, but *caveat emptor*! (Let the buyer beware!) – for reasons that will shortly be made clear (see 'Using contractors – the downside').

Professionalism

As mentioned earlier, many contractors concentrate on doing one or two activities and in consequence develop a very high quality service.

Specialism

Some work is so specialized, complex or hazardous that there is no alternative but to use a specialist contractor or vendor.

Productivity and cost

Audits indicate that good contractors provide higher productivity at lower cost than the client could achieve with its own labour. Also, it is easier to replace sub-standard contract workers.

Using contractors – the downside

There are also disadvantages to using contractors.

Unavailability when required

Contractors will also be required by other companies for similar work and may not be available when required. Dependence on contractors may therefore give rise to problems if the procurement system is not of the highest calibre.

Contractors not always as effective as their proposals suggest

One of the reasons for this arises out of the above unavailability problem. If the contractor is successful his resources may become stretched. If he cannot cover the contract with his normal personnel he has a problem. Because he doesn't want to lose the work he may provide some of his key staff and then supplement them with people 'off the street' who are of unknown or variable quality – hence the earlier *caveat emptor*.

One factor should be borne in mind. The contractor's proposal will invariably have been based on the performance of a full strength team. If he fields a second string team the client doesn't get the quality paid for.

Safety performance

If the contractor's safety systems are sub-standard, or if the men provided by the contractor are not well-practised at working on hazardous sites, the accident rate can increase. The safety performance of contractors is, in some cases, not comparable with that of the client (although, it must be said, since the introduction of SDM legislation the reverse is often true).

Difficult client–contractor relationship

A difficult relationship may end, in extreme cases, in a conflict that can only be resolved via litigation, which usually centres on claims for payments for delays, changes of intent and emergent work, but can also be due to the fact that the client or the contractor (or both) underestimated the volume or difficulty of the work.

Conflict between contractors

If many contractors are used (on major Turnarounds they could number more than twenty) there will be issues to resolve at the interfaces between them – the most common are:

- One contractor delaying another.
- Two contractors trying to work in the same area at the same time.
- One contractor undoing another one's work.

For example, the welding contractor completes a number of welds on a pipe by eight o'clock in the evening. Radiography has been booked overnight. The following morning the lagging contractor (in conformance with the schedule) insulates the pipe. Later that day he finds the lagging has been stripped off. Due to a hold-up on another job the NDT team hadn't radiographed the welds overnight and the insulation had to be removed to allow the late radiography of the welds. Whose fault is it? Who should pay for the re-work? These are problems of co-ordination and are ultimately the responsibility of the Turnaround manager.

Agency labour (body shops)

One difficulty which can constitute a *force majeure* may arise if the contractor hires mainly agency labour. Because Turnarounds are of short duration and therefore often regarded as 'casual work' agency labour will always be on the lookout for new work. If there is the offer of longer or more lucrative employment at another location, such manpower may quit the Turnaround without warning and head for greener pastures, leaving the client with insufficient labour to complete the Turnaround within the planned duration – and in the short term there is nothing that can be done about it. The company may be especially vulnerable when the labour which disappears is a scarce resource such as coded welders or instrument technicians.

It is perhaps stating the obvious to say that care and attention is required in the creation of contractor packages and the selection of contractors. Experience teaches most Turnaround managers that this is a difficult and demanding task and the message needs to be constantly re-emphasized to the Turnaround team. Using contractors will not solve all the company's problems. Rather, it will exchange one set of problems for another.

Contractor work packages

The main factors influencing the selection of contractors are:

- the work scope and how it is packaged;
- the design of the Turnaround organization;
- the type of contract to be awarded;
- the availability of contractors.

Because they are inter-connected they also influence each other.

There are a number of specific options for packaging work, i.e. by work type, functional unit, geographical area or contractor availability. In reality, the most likely situation will be that a combination of the above options is integrated into an overall contractor plan.

By work type

A given work package is created by grouping similar tasks together and awarding the whole to one contractor to make it more manageable and more economic to perform. The most common packages are as follows (a representative but not exhaustive list):

- Valve overhaul and replacement;
- Pump overhaul and replacement;
- Pressure vessel inspection;
- Re-traying columns;
- Catalyst and packing;
- Machines overhaul;
- Instrumentation;
- Electrical work;

- Welding (especially coded welding);
- Water washing and cleaning;
- Scaffolding (or staging);
- Lagging (or insulation);
- Painting (or protective coating).

The well-defined package forms the work scope for the contractor. Because each package is made up of similar repetitive tasks it is simpler to price, and its execution easier to control. On the other hand, the resulting engagement of a number of contractors can give rise to contractor–contractor conflict.

By geographical area

The plant is divided into discrete areas and all of the work (apart from highly specialized or hazardous tasks) in the area is given to one main contractor. While this arrangement can be complicated, due to the mixture of different tasks that the one contractor must perform, the advantage is that, having only one contractor in the particular area, there is no need to manage issues between contractors.

By functional unit

One contractor is awarded responsibility for a whole system (steam, for example,) undertaking many different tasks across a number of geographical areas. The obvious advantage is a greater guarantee of resulting system integrity because it is under a unified control. Conversely, if several contractors work on different parts of one functional system the Turnaround manager must provide a co-ordinator to ensure all interfaces between contractors are properly managed and no work is left undone (because one contractor mistakenly believes that one of his tasks is part of another's work package).

Contractor availability

Contractor availability cannot be guaranteed. It is unwise to take it for granted and to expect the contractor to be available – on short notice – just when you need him. Availability depends upon a contractor's workload and this in turn depends upon a number of factors such as:

- the time of year (Turnarounds occur cyclically – fewer are done in winter);
- the concentration of contractor dependent clients in the particular area;
- the type of work involved (the more specialized, the fewer people do it);
- the amount of work involved (major events may require thousands of men).

A Turnaround manager has to protect the interests of the client. Giving due regard to the above factors, he must adopt a strategy, which would include, *inter alia*, the following elements:

1. Finding out when other companies in the area will be undertaking Turnarounds;
2. Identifying windows of opportunity to execute the overhaul;
3. Compiling a list of contractors and auditing their suitability;
4. Determining the availability of suitable contractors at the time of the Turnaround;
5. Inviting competitive tenders. (Sending out invitations to bid (ITBs) as early as possible);
6. Booking specialist contractors as early as possible.

and, most importantly, securing the future by building long-term relationships with high quality contractors.

Types of contract (see Figure 5.1)

There are a number of different types of contract which can be awarded. The award will depend upon the circumstances surrounding particular packages of work. The types of contract are defined separately below, but in reality a complex project such as a Turnaround will consist of a combination of any (or all) of the options shown. They are:

- Single contractor managed contract;
- Management fee and reimbursed labour cost;
- Fixed-price packages;
- Call-off contracts (on scheduled rates);
- Day-work rates.

Each of these types of contract has its own features, benefits and drawbacks (see Figure 5.1).

When awarding contracts there must be a clear understanding of all the circumstances surrounding the Turnaround from the perspective of:

- Contractor resource availability;
- Time-frames and windows of opportunity;
- Available finance;
- Different types of work required;
- Complexity of work;
- Need for specialized tasks;
- Ability to manage the different types of contract.

Once the types of contract have been chosen, the next step is to select the contractor.

Check list of issues to be considered

1. Single contractor managed contract

- *Features*
  - Requires a well defined, validated work scope
  - One contractor is awarded the total work scope
  - Contractor bids a fixed price for the contract
  - Contractor plans and co-ordinates all work
  - Contractor may execute work or sub-contract specific packages to other companies
  - Client facilitates contractor effort and monitors progress and performance
- *Benefits*
  - Contractor manages the total event
  - Contract cost and duration are fixed
  - Contractor must conform to agreement
  - Contract variations can be handled easily
- *Drawbacks*
  - Expensive preparation period required
  - Client needs excellent commercial skills
  - Client pays for all delays and downtime
  - Contractor's may underbid to win the contract and then try to generate extra work
  - If duration is tight, contractor may cut corners and compromise safety and quality
  - Client and contractor may become adversaries

2. Management fee and reimbursement

- *Features*
  - Contract agreed before work scope finalized
  - Contract executed to a fixed duration
  - Contractor manages and plans the event and provides labour to carry out all work
  - Contractor may subcontract work packages
  - Client pays a fixed fee for management and reimburses contractor for labour hours expended
  - Client monitors and validates contractor man-hours
- *Benefits*
  - Does not require early validation of work
  - Rates offered by different contractors can be compared and the best deal negotiated
  - Client can negotiate productivity agreements
  - Contractor bears the cost of over-run
- *Drawbacks*
  - Encourages slack planning by the client
  - Total cost will not be known until after the event
  - Client must closely monitor contractor man-hour claims to avoid 'padding out'
  - Work content cannot be controlled as well as on a fixed price contract
  - Work may overwhelm resources

3. Fixed price packaged contracts

- *Features*
  - Similar work grouped into a package
  - Contractor bids a fixed price for the package
  - Contractor manages the package
  - Client manages the contractors
- *Benefits*
  - Costs and duration are fixed
  - Onus on contractor to work effectively
  - Usually very few variations
- *Drawbacks*
  - Client pays for all delays and downtime
  - A number of contractors whose needs conflict
  - Client needs excellent co-ordinating skills

4. Call off contracts (scheduled rates)

- *Features*
  - Used for work that is difficult to specify exactly but can be categorized (scaffold etc.)
  - Works cope does not need to be defined
  - Client manages the contractors
- *Benefits*
  - Once agreed, all tasks are done at unit cost
  - Variations handled at unit cost
  - Client can negotiate productivity deals
- *Drawbacks*
  - Requires quantity surveying by client
  - Call off system is relatively easy to abuse
  - Can generate daywork with resulting costs

5. Daywork rates contract

- *Features*
  - Manpower hired to perform unspecified work
  - Contractor or client may supervise
  - Usually for general work only
- *Benefits*
  - No planning required – short lead time
  - Manpower more readily available
  - Easier to vary numbers at short notice
- *Drawbacks*
  - Very variable skill level – usually low
  - Client has to manage labour – high costs
  - Tend to be hidden costs and expenses
  - Difficult to achieve good productivity

Other considerations

Figure 5.1 Types of contract

Contractor selection (see Figure 5.2)

In order to maximize the likelihood of the contractor performing effectively, the Turnaround manager should consider the following actions:

Before the event

- If possible, use a known, high quality, contractor. (If contractors are unknown quantities, invite at least three to bid for each package.)
- Draw up a short list of companies and if possible, audit and compare them.
- Send invitations to bid (ITBs) out early to give the contractors sufficient time to assemble their bids.
- In all ITBs, clearly specify:
  - an accurate work scope (what is required, to what standard);
  - any special requirements (e.g. the need for qualified coded welders);
  - any special circumstances (e.g. working next to live, hot plant);
  - any built-in non-productive time (e.g. awaiting permits to work).
- When the contractor's bid are received, evaluate them and negotiate improvements to unsatisfactory elements.
- Select the contractor most suited to the needs of the Turnaround.
- Agree achievable, measurable targets with the contractor – on safety, duration, hours expended, quality, or any other useful indicator.
- Ensure the contractors' IT systems are compatible with the client's systems or that there is an effective translation system.
- Ensure that, if the contractor is planning the overhaul work, the plant shutdown–start-up network can be linked to the contractor's schedule.

Factors influencing the selection of contractors

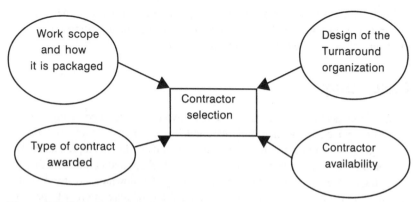

Figure 5.2 Selection of contractors

- Place the order for the contract as early as possible, to give the contractor time to prepare for the contract.
- Prior to the start of the event, thoroughly brief the contractor's employees on the standards required.
- Ensure that all necessary documentation is issued before the time it is required.
- Ensure that skill levels are as promised by the contractor.

During the event

- Closely monitor the contractor's performance against the agreed targets and ensure steps are taken to rectify any sub-standard or late work.
- Ensure the client organization does nothing to hold the contractor up (the most important matter here is the timely issuing of permits to work).
- Minimize knock-on effects from one contractor to the next by effective co-ordination of the total task.
- Speedily resolve all client–contractor and contractor–contractor conflicts.
- Ensure all variation instructions and emergent work requests are approved and issued in written form.
- Ensure contractors understand that they will not be paid for any work they have carried out on a verbal instruction.
- Validate contractor claims at the earliest opportunity.
- Sign off completed contractor work as early as possible.

After the event

- Thoroughly debrief the contractor to establish 'lessons learned'.
- Score contractor's performance and retain for future reference.
- Develop improved contractor programmes for future events.
- Gather and quantify the history of the overhaul (extremely useful data for the planning of future Turnarounds).

Contract management is becoming a larger part of Turnaround management as companies rationalize, divest and downsize. The emphasis is therefore moving from managing the work to managing contracts.

Case study

A client contracted a very large and well-known construction company to schedule and execute and overhaul a Turnaround. The client had already planned out the majority of individual jobs on method sheets using Lotus Approach. Before the contract was signed it was assumed (by both client and contractor) that the large data base of information could simply be downloaded from Lotus Approach to the contractor's Primavera scheduling software in order to produce the schedule for the Turnaround. In the event, the IT representatives from both companies found it impossible to do this and the result was that after much agonizing and negotiation between client and contractor, it was agreed that the latter would take on extra staff (at the client's expense) to manually key in the data which was read from hard copies of the Lotus Approach reports.

The client checked to ascertain if any difficulties would be experienced in linking the plant shutdown and start-up networks to the contractor's schedule. It was discovered that, for this task, the client had used Primavera software and it was agreed this would be compatible with the contractor's software – also PrimaVera.

Unfortunately, when the time came to link the networks to the schedule it was found that, although both programmes were on Primavera, the client had a later version of the software and the link up was never made.

Once again the data had to be input manually. This meant even more staff had to be put on planning (again at the client's expense). A far greater consequence was that the Turnaround schedule was not ready for the start of the event. The critical path, and therefore the duration of the event, was unknown. From day one the Turnaround was out of the manager's control.

What could have been done to avoid this?

6

Planning the Turnaround

Introduction

In the introduction it was stated that a Turnaround was unique because there were differences between Turnarounds and other types of project. One of the differences is the approach used to plan and schedule Turnarounds and other projects. In fact it may not be going too far to say that Turnaround planning is performed in the reverse order to other projects (see Figure 6.1).

The contrast between Turnarounds and other projects

A normal project starts off with the project definition (e.g. let's build a bridge). Then, using a technique called 'work breakdown' the whole project is broken down into smaller more manageable chunks (e.g. fabricate bridge structure, dig foundations etc.) and then into even smaller chunks (e.g. for bridge fabrication – formulate a design > procure materials > set up a fabrication facility > fabricate bridge etc.), the chunks are then scheduled on a timeline and then broken down into individual tasks (e.g. for fabrication – cut steel tubes to size > prep for welding > weld tubes together etc.) > radiograph welds etc.).

Figure 6.1 Contrasting project and Turnaround

Each individual task is generated by the need to fulfil the larger chunks of work and each chunk is generated by the need to complete the total project.

By contrast, when planning a Turnaround we start off with an undifferentiated work list of many varied individual tasks that come from a number of different sources – all of them unconnected. We validate the tasks (justify > clarify > categorize) and then, using the reverse technique of 'work buildup', we plan each of the individual tasks and integrate them into a Turnaround schedule from the individual tasks. Only when we have completed these steps can we define the project in any credible way.

This may be one of the reasons why a number of experienced project managers (including the author) struggle when managing their first Turnaround.

A Turnaround plan may be formulated over a long period of time in preparation for a large amount of work which is going to be performed in a very short space of time. It is not uncommon to plan for nine months a Turnaround which will be completed, product to product, in less than three weeks. The normal Turnaround features a high volume of work carried out by a large number of people working under extreme time and access constraints. It therefore requires planning of an order and detail that may not be necessary during normal production.

The basic objective of planning is to ensure that the right job is done at the right time by the right people. Figure 6.2 shows how this objective is met using work specification, work scheduling and resource scheduling.

The planning of a Turnaround requires the active participation of a large number of people including, among others, the following:

1. *Preparation team*: under the leadership of the Turnaround manager create the plan which will be presented to the steering group.
2. *Plant team*: provide basic data, work requests, technical information and the shutdown–start-up network and then validate final planning documentation.

Planning strives to ensure that the right job is done at the right time by the right people, therefore the main activities are:

The planning of a turnaround requires the active co-operation of many people, including:

- Preparation team
- Plant personnel
- Inspectors
- Engineering department
- Project managers and engineers
- Contractor's representatives
- Suppliers' and vendors' agents
- Equipment manufacturers

Production of the master plan is the responsibility of the turnaround manager and preparation team

Figure 6.2 The basic objective

3. *Inspectors*: specify inspection work, requirements and techniques.
4. *Engineering department*: provide technical information and support on a range of topics.
5. *Project managers and engineers*: provide the planning and documentation for project work.
6. *Contractor representatives*: advise on feasibility of their part of the plan.
7. *Steering group*: approve and fund the final plan.

The main elements in creating a Turnaround plan are shown in Figure 6.3 and demonstrate the necessity for co-operation from all those concerned.

The first stage in planning is the creation (at a series of 'Work List Review Meetings') of task specifications, considered under three basic headings – major tasks, small tasks and bulkwork – a different planning method being adopted in each case.

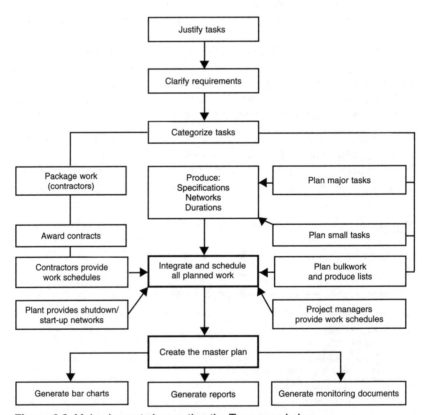

Figure 6.3 Main elements in creating the Turnaround plan

Major tasks (see Figure 6.4)

A major task is one that fulfils one or more of the following criteria:

- it is abnormally hazardous;
- it is technically complex or unfamiliar;
- it has a high work content;
- it will involve a large number of people with different skills;
- it involves multi-level working;
- it will take a long time to complete.

Check list of requirements

1. Major task method statement A detailed document which defines the following: – step-by-step procedure for performing the task – list of fully specified materials – day-by-day profile of manpower requirements – day-by-day schedule of tools and equipment – day-by-day schedule for services requirements – day-by-day schedule of utilities requirements – notification of any special circumstances or requirements	**2. Task networks** A flowchart which shows the following: – each task activity in logical sequence – a duration for each activity – the 'longest time' critical path – any allowance for contingency – slack time for shorter activities
3. Isolation register A document which contains: – a sketch or list defining the position of every isolation required to make the task safe – the type of isolation at each position, e.g., slip plate, spectacle plate, physical disconnection – a detailed schedule of the isolation plate ratings – a certificate which the plant responsible person must sign to indicate isolation is satisfactory	**4. Welding procedure** A document which contains: – a sketch of the butt to be welded – parent metal and electrode material specifications – weld process to be used – pre and post weld heating requirements – rectifier type and weld current – weld test required to qualify welders – none destructive test requirements
5. NDT request and report A document which details: – the type of non-destructive test required – the date and time the test is required – a report sheet for detailing the results of the test	**6. Pressure test request and report** A document which details: – plant item number/line number – test medium and temperature – test pressure and duration – sketch of feed and bleed points – type of equipment and pressure gauge number – pressure and temperature recorder numbers – pump details – name of tester and acceptor – results of test, and comments
7. Task safety plan A document containing: – a task hazard report – a rescue plan	**8. Any other necessary documents** Any documents needed to clarify or assist the performance of the task, e.g.: – statutory documents – assembly or detail drawings – standards or specifications – plant SOPs – manufacturer's or vendor's recommendations

Figure 6.4 Major tasks

Some examples of major tasks are:

- overhauling a large machine;
- replacing all of the elements of a long conveyor belt;
- applying an internal protective coating to a large vessel with a small manway access;
- demolishing a redundant cooling tower;
- installing a new computerized instrumentation control system (such as a DCS);
- re-traying a large distillation column.

Such tasks may require the input of an engineer. In the case of Turnarounds it is the preparations co-ordinator who works up the work packages for the major tasks. His job is to transform the basic data into a set of written instructions and supporting documentation. This provides sufficient information to enable someone to carry out the task with as little reference to outside sources as possible – which means that the major task work-up covers *every* known aspect of the task (see Chapter 3, where the major task review meeting was discussed in detail).

The following information package is created by the review team led by the preparations co-ordinator (see Figure 6.4 for a detailed listing of the requirements of the information package):

1. Statement of major task method.
2. Task network.
3. Isolation register.
4. Welding procedure (if required).
5. NDT request and report forms.
6. Pressure test procedure and report forms (if required).
7. Safety plan (see Chapter 10).
8. Any other necessary documents.

When the task package is complete the preparations co-ordinator passes it to the planning co-ordinator for inclusion in the Turnaround schedule. At a later date, usually two to three weeks before the start of the event, the preparations co-ordinator will present the major task packages to the steering group, and other interested personnel, at the final review meeting for major tasks.

Minor tasks (see Figure 6.5)

Minor or small tasks do not normally require the input of an engineer but they do require individual specification by an experienced planner. Typical such tasks are:

- clean, inspect and if necessary repair a medium size or large heat exchanger;
- inspect and if necessary repair a small vessel;

Check list of requirements

1. Small task method sheet

A standard form which specifies the following:

- Order number and task number
- Tag number and location of task
- The job method
- A materials list
- Manpower required
- Tools and equipment required
- Services required
- Utilities required
- Any special requirements

2. Small task networks

A simple flowchart cross-referenced to the task method sheet, showing the following:

- The main steps of the task in logical sequence
- A duration for each main step
- An overall duration
- Any contingencies

3. Isolation register

A standard form, cross-referenced to the task method sheet, specifying:

- A sketch or list defining the position of every isolation needed to make the task safe
- The type of isolation at each position, e.g., slip plate, spectacle plate, physical disconnection
- A detailed list of isolation plate ratings
- A certificate which a responsible plant person must sign to indicate isolation is satisfactory

4. Welding procedure

A standard form, cross-referenced to the task method sheet, specifying:

- A sketch of the butt to be welded
- Parent metal and electrode specifications
- Weld process to be used
- Pre and post weld heating requirements
- Welding rectifier type and weld current
- Weld test required to qualify welders
- None destructive test requirements

5. NDT request and report

A standard form, cross-referenced to the task method sheet, specifying:

- The type of NDT task required
- The date and time the test is required
- A report sheet detailing the results of the test

6. Pressure test request and report

A standard form cross-referenced to the task method sheet, specifying:

- Plant item number/line number
- Test medium and temperature
- Test pressure and duration of test
- Sketch of feed and bleed points
- Type of equipment and pressure gauge number
- Pressure and temperature recorder numbers
- Pump details
- Name of tester and acceptor
- Results of test, and comments

Other considerations

Figure 6.5 Small tasks

- replace a nest of twenty-four steam traps;
- remove, overhaul and replace a large pump.

The planner produces, as necessary, a combination of some or all of the following documentation (Figure 6.5 gives a more detailed listing of the requirements of the information package):

- Task method sheet;
- Small task network;
- Isolation register;
- Welding procedure;
- NDT request and report form;
- Pressure test procedure and report form.

Once the planner has completed the small task package he passes a copy of it to the planning co-ordinator for inclusion in the Turnaround schedule. He retains the original that will later be bound – with other relevant task packages – into an area, unit or work-type planning book for issue to the supervisor who will execute the work.

The work pack

In order for the supervisors to organize work and tradesmen to carry it out they need a package that contains all of the information required for the task. This allows for rational execution of the task and saves time being wasted on the day, hunting for information. We will now review some of the documents in the work pack.

Job control sheet

Figure 6.6 is an example of a common type of job control sheet used in the UK. It was originally designed so that it could be used by someone who had never been on the plant before – to cover all eventualities. It has a number of interesting features.

Plot plan

This is a map to show the person doing the task where on the plant the job location is.

Digital photograph

This is a photograph of the actual item of equipment to be worked on so that the person can recognize it when he reaches the location.

JOB TITLE 3 YEARLY OVERHAUL OF P132 Pump Assembly				REVISION 2	
☐ SCAFFOLD REQ ☒ CLEANING CERT ☐ LIGHTS REQ ☐ ☐ WQP REQ ☐ ☐ LAGGING TO REMOVE ☐ ANTI GAS MAN REQ ☒ ENTRY REQ ☐					
JOB ACTIVITY STEPS					
1	Collect Permit 4424 & carry out Take Time - (check isolations)	F			
2	Collect permit 4424E isolate and disconnect P32 electrics	E	1		
3	Disconnect P132	F	3		
4	Rig P132 and load onto forklift	R	1.5		
5	Transport P132 to cleaning bay and unload	D	0.5		
6	Sign off Permit 4424 and hand back	F			
7	Decontaminate P132	C	0.5		
8	Inspect and issue cleaning certificate	I	0.5		
9	Load P132 onto wagon and transport to Workshop B1	D	1.5		
10	Check decontamination certificate	WS			
11	Carry out 3 yearly PMR on P123 as per MM1004	WF WE	12		
12	Load P132 onto wagon and transport to clean lay-down area	D	1.5		
13	Collect P132 with forklift and transport to site	D	0.5		
14	Collect Permit 4661 and carry out take time	F			
15	Rig P132 and lift into position	F, R	1.5		
16	Reconnect P132 (use Spiral Gasket and new fastenings)	F	4		
17	Sign off Permit 4661 and hand back	F			
18	Reconnect P132 electrics and de-isolate	E	1		
19	Sign of Permit 4424E and hand back				
PLANT S & D	AREA Export	SYSTEM Product		PLANNER A Brightman	
BERTH/ PARENT ID ST132		EQUIPMENT No P132		EQUIPMENT NAME Product Export	
JOB No T229	DRG No N/ A	REG P/WORK No L014		ELECTRICAL ISOLATION	
FITTER'S ID			WORK COMPLETE	OPS ACCEPTANCE	

Figure 6.6 Job control sheet

Services suite

At the top left of the sheet there are a number of tick boxes where the planner records his service requirements. If any box is ticked then a copy of the sheet goes to the person responsible for providing that service to allow them to plan for it.

Job activity steps

This is what most people think of as planning, a sequence of every activity required to complete the task (often referred to as 'natural job steps'), written out to allow the job to be carried out in a logical manner from start to finish. It includes all permit requirements (the job number is also cross-referenced on the permit) and, in the bottom right-hand corner of the sheet, specifies the type of isolation required ('pull the fuses').

The job activity steps are best formulated by the planner visiting the site and mentally 'walking through the job', noting any constraints or difficulties that may be experienced and then tabulating the activities – usually on a computer.

Also notice in activity one the instruction 'carry out Take Time'. This refers to a safety inspection that is required to be performed by the work crew immediately before starting work to ensure that the job area is safe and nothing has been changed since the permit was signed (which may have been some hours before).

Trades

This column specifies the trades that will be required for each activity (F = Fitter, E = Electrician etc.).

Hours (HRS)

This column specifies the time allowed to carry out each activity. If there is more than one person in the Trades column then the time is multiplied by the number of people to derive the total hours for the job.

Technical information

This is supplied at the bottom of the sheet and represents all of the information required by the various people involved in the work. The bottom line is interesting. The first box (FITTER'S ID) is there because there are flanged joints to be broken and remade on this particular job and, in this company, every fitter must 'sign' for the joints he makes.

The reason for this is that a number of years ago there were a significant number of leaks when the plant was started up after a Turnaround and, when the Turnaround manager tracked down who had made which joint by interviewing the supervisors and getting them to remember which fitters had been allocated which work, they found that the vast majority of the leaking joints had been made by only two fitters. So they added this box to the sheet to make identification simpler in the future.

The second box on the bottom line requires the supervisor to sign off the job as complete (single point responsibility at work!) and the third box requires a member of the operations staff to check the work done and sign it off as acceptable.

The job control sheet has two great advantages:

1. It covers all aspects of the work and forms a complete set of instructions – all on one sheet (or maybe two for larger jobs);
2. Once it has been created it can be reviewed and used every time the job is done – saving many planning hours in the future.

Isolation register sheet

Figure 6.7 is an example of another important planning document in the work pack – the isolation register sheet. The example is from the chemical industry and shows the most extreme form of physical isolation, known as 'Physical Disconnection' or just PD.

This involves removing sections of pipe work (known as spools or bobbins) from all pipelines connected to the equipment so that there is 'fresh air' between the equipment and any source of energy or hazardous fluid.

Isolation Register Sheet (Physical Disconnection)							
JOB DESCRIPTION: Pressure Vessel Inspection						PLANT: Caustic Chlorine	
Work order No 177	EQUIPMENT: Cooled Caustic Chlorine		EQUIPMENT No T179			AREA: B	
No	BRANCH DUTY	SIZE	STUD	GASKET	ISOLATION REQUIRED	TAG
A	Overflow to T720	200/150	L7/L4	M228	Remove Section of line	700
B	Filter Recycle	150/150	L7/L4	M228	Remove bobbin & blank line	701
C	Filter Recycle	150/150	L7/L4	CAF	Remove bobbin & blank line	702
D	Cooler Exit line	150/150	L7/L4	CAF	Remove bend & blank line	703
E	Level LT7638	75/150	L7/L4	M228	Disconnect & remove valve	704
F	ExP727 Suction Line	150/150	L7/L4	CAF	Remove bobbin & blank line (BOUNDARY ISOLATION)	102
G	P723 Suction Line	150/150	L7/L4	CAF	Remove bobbin & blank pump	705
H	P719A Suction Line	150/150	L7/L4	CAF	Remove bobbin & blank pump	706
J	P19B Suction Line	150/150	L7/L4	CAF	Remove bobbin & blank pump	707
K	Manhole	600 Sq.	L7/L4	M228	Remove Manhole Cover	
L	Vent	N/A	N/A	N/A	No action required	
M	Agitator	N/A	N/A	N/A	Electrically Isolate and remove drive belts	

NOTE – ALL LINES NEED TO BE SUPPORTED BEFORE JOINTS ARE BROKEN

SCAFFOLD REQUIRED TO POINTS: INSULATION REMOVAL FROM POINTS:

A at each end. Around tank for Wash.   F /G /H /J

Board out fragile roof

Figure 6.7 Example of an isolation register sheet

The sheet supplies all of the information needed to safely isolate the equipment. One thing of particular note for shutdown planning – isolation 'F' is identified as a 'Boundary Isolation', which indicates that it will form part of the isolation for more than one task. It is extremely important that all tasks involved are cross-referenced and that this isolation is not removed until all the tasks are completed.

Job safety analysis

In order to plan safely, it is necessary to understand the hazards associated with the task. Figure 6.8 shows an example of a simple, but very effective technique to identify, evaluate and deal with the hazards. The process has four activities:

1. Write down each of the natural job steps involved in the task;
2. For each job step, identify any hazard associated with it – this will normally require a visit to the plant to view the job site;
3. Define the potential loss to people or the environment associated with the hazard;
4. Specify precaution to be taken to either eliminate the hazard or, if that is not possible, to guard against it.

Using this technique effectively in planning work reduces the risk of accident to a minimum.

Job Number...*193*......Job Title...Overhaul isolation Valve ...*IV1222*.........Date...03.03.04

Job Step	Associated Hazards	Potential Loss	Precaution
Remove section of walkway grating for access to valve.	Hole in walkway.	Person could fall through hole and sustain serious injury.	Erect barrier around hole and post 'Hazard – Keep Out' signs on both sides of the walkway.
Disconnect valve.	Men working at height on scaffold platform.	Items could fall from platform and Injure people working below.	Erect safety net underneath the scaffold platform and post 'Men Working Above' signs at ground level.
Rig valve and lift with 350te crane onto wagon.	Valve is swung over area where men are working at ground level.	Valve could fall on people causing severe injury or death.	Stop work and clear area of people while valve is being loaded onto wagon. Sound alarm siren.

Figure 6.8 Job safety analysis (risk management)

Other documents required to carry out the task in the work pack include but would not be limited to:

- BOM (bill of materials)
  The planner specifies all of the materials needed to perform the task. In some companies the material is broken down into four 'bins'. Not all bins are required for all tasks. They are:
  Bin 1 – Materials required to isolate the equipment
  Bin 2 – Materials required to carry out the task (spares etc.)
  Bin 3 – Materials required for testing (e.g. pressure testing)
  Bin 4 – Materials required for de-isolating and boxing up
- Drawings;
- Technical specifications;
- Line diagrams;
- Pressure test procedures and test sheets;
- Radiography request forms;
- Inspection request forms.

The work pack allows the supervisor to:

- 'Walk through' the job activity steps before the event to ensure that they are feasible;
- Identify interfaces with other work in his area and take action to prevent any potential hazards;

- Analyse logistical requirements and organize all aspects of the job;
- Report any errors in the work pack to the planner before the event so that they can be corrected;
- Ensure that all documents, materials, equipment and services required for the task will be available on the day.

Bulkwork (see Figures 6.9 and 6.10)

Bulkwork consists of large groups of identical, or similar, simple tasks that do not need to be specified in individual task sheets but can be bulked together on one list. The planner produces the list of work, typically covering the overhaul or replacement of such items as:

- valves (control, relief, isolation, return etc.);
- small pumps;
- bursting discs;
- orifice plates;
- other simple units.

Network plans are not needed for these jobs because they are normally included in the schedule as blocks of work. This is done purely to derive a meaningful resource profile, because existing software scheduling is not ideal for handling large numbers of short duration tasks. The scheduler will use the short duration tasks individually to fill holes in the resource requirement profile caused by the quiet periods on major jobs. Consider, for example, a relief valve and a control valve adjacent to each other on top of a thirty metre high column. To fill resource gaps, the schedule may programme removal of each valve on a different day. This is clearly a waste of time and resource. Therefore, valves and other small items are normally scheduled and marshalled manually. Figure 6.11 shows a manual valve marshalling sheet which defines the number of valves to be removed each day and the order in which they are to be decontaminated, sent for overhaul and returned to site.

Thus far, it has been assumed that tasks fall neatly into one of the three categories discussed, viz major, small, and bulkwork. What is to be done, however, with a task that falls on the border line between two categories? Remember that a Turnaround is a hazardous event – take no risks! If there is any doubt whatsoever, always elevate the task to the higher category.

The list of tasks provides the raw data for defining what is sometimes referred to as the 'mechanical duration' of the Turnaround, i.e. all tasks that are carried out between the time the plant (or any relevant section of it) is shut down and the time it is started up. The additional information that is needed to produce a schedule for the total event is termed the 'shutdown–start-up logic' (SDSU).

YEAR............ PLANT............................ UNIT / AREA..................... ITEMS..................... No: ....

PLANT ITEM NUMBER	OVERHAUL OR FIT SPARE?	ASSOCIATED PLANT ITEM	TAG NUMBER	GRID REFERENCE	HEIGHT FROM GROUND IN METRES	WEIGHT OF ITEM IN KG	SCAFFOLDING REQUIRED	DELAGGING REQUIRED (ASBESTOS ?)	CRANE SIZE REQUIRED	MECH.	ELEC.	INSTR.	CIVIL

Figure 6.9 Bulkwork specification sheet

YEAR............ PLANT.................... UNIT / AREA.............. ITEMS.............. No: ....

ITEM NUMBER	ASSOCIATED PLANT NUMBER	INST / ELEC DISCONNECTION (IF REQUIRED)	REMOVED	DECONTAM.	SENT TO WORKSHOP FOR OVERHAUL	RETURNED FROM WORKSHOP	OVERHAULED ITEM FITTED	SPARE ITEM FITTED	INST / ELEC RECONNECTED (IF REQUIRED)	TESTED AND COMMISSIONED

Figure 6.10 Bulkwork control sheet

Figure 6.11 Valve marshalling sheet

Dispatch to workshops					Latest acceptable return from workshops					
DAY 2	DAY 3	DAY 4	DAY 5		DAY 23	DAY 24	DAY 25	DAY 26	DAY 27	DAY 28
CV1001	CV1012	CV1023	CV1234		CVIOOI	CV1102	CV1013	CV1023	CV1234	RV132
CV1O22	CV1013	CV1006	RV132		CVIO22	CV1209	CV1014	CV1006	IV1108	RV133
CV11O2	CV1014	CV1007	RV133		RVIO5	CV1012	VR108	CV1007	IV1109	IV1200
CV1209	RV108	CVI2O4			RV106	RV101	RV110	CV1204		
RV105	RV110	CV1205			IV1006	RV109	RV102	RV123		
RV1O1	RV102	RV122			IV1007	RV189	RV122	RV124		
RV109	IV1012	RV123					IV1012	RV125		
RV189	IV1008	RV124					IV1008	IV1102		
IV1006	1V1009	RV125					IV1009			
IV1007		IV1102								
		IV1108								
		IV1109								
		IV1200								

Sequence determined by the needs of the schedule and by accessibility of the items

The shutdown–start-up logic

Options (see Figures 6.12 and 6.13)

The combining of the shutdown–start-up logic with the mechanical duration logic will have a fundamental effect on the planning and execution of the Turnaround. There are four main options and each will involve the Turnaround manager in a different role with different responsibilities. The options are listed below (also see Figure 6.12) and the features, benefits and drawbacks of each are defined in Figure 6.13.

1. The Turnaround manager is responsible for the mechanical duration only and the plant manager for shutdown and start-up.
2. The Turnaround manager is responsible for the mechanical duration, the plant manager for shutdown and start-up with Turnaround manager acting as a shutdown–start-up consultant.
3. The Turnaround and plant managers combine as a team for the total event with a more senior manager having overall responsibility.
4. The Turnaround manager is responsible for the total event.

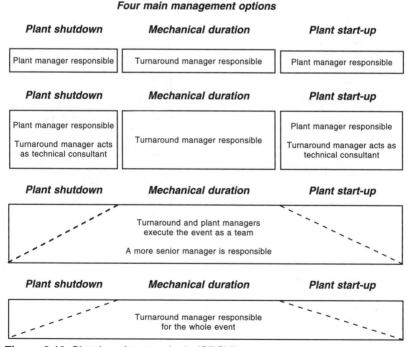

Figure 6.12 Shutdown/start-up logic (SDSU)

1. Mechanical duration only

- *Features (Separate steps)*
  - Plant staff shut the plant down, decontaminate and cool it then hand it over to turnaround team
  - Turnaround team complete all planned work and hand the plant back to plant staff
  - Plant staff start the plant up assisted by a start-up team supplied by the turnaround manager
  - Well defined cut-off points
  - Plant manager is responsible for shutdown/start-up
  - Turnaround manager is responsible only for the mechanical duration
- *Benefits*
  - Turnaround planning simpler and cheaper
  - No formal shutdown/start-up network required
  - Plant may be shutdown/started up in any order
  - Common start for manpower
  - Overrun costs charged to the plant
  - Risk of accidents/incidents less than Options 3 and 4 but greater than Option 2
- *Drawbacks*
  - Product to product duration will almost certainly be longer
  - Production loss will almost certainly be higher

These drawbacks may not apply if the critical path task has the longest shutdown/start-up duration

2. Mechanical duration + consultancy

- *Features (Separate steps)*
  - Turnaround team advise on best practice for shutdown/start-up programmes
  - Plant staff shut the plant down, decontaminate and cool it then hand it over to turnaround team
  - Turnaround team complete all planned work and hand the plant back to plant staff
  - Plant staff start the plant up assisted by a start-up team supplied by the turnaround manager
  - Well defined cut-off points
  - Plant manager is responsible for shut-down/start-up
  - Turnaround manager is responsible only for the mechanical duration
- *Benefits*
  - Duration may be shorter than in Option 1
  - Production loss may be less than in Option 1
  - Custom and practice is challenged to find more effective ways of shutting down and starting up
  - SDSU overrun costs charged to the plant
  - Less risk of accidents/incidents than other options
- *Drawbacks*
  - Turnaround team workload is increased
  - A formal shutdown/start-up network is required
  - If network is not adhered to, work may be delayed

3. Integrated plant/turnaround team

- *Features*
  - Plant/turnaround staff form one team and produce an integrated turnaround plan, product to product
  - The plant is shut down, decontaminated and cooled system by system, in a pre-planned sequence
  - Equipment is opened up and worked on as soon as safe, while other systems are being shut down
  - Systems brought on line in a planned sequence while other systems are still being overhauled
  - Turnaround and plant manager manage jointly

A senior manager has overall responsibility
- *Benefits*
  - Product to product duration can be significantly shorter and product loss less than Options 1 and 2
  - A mixture of plant team local knowledge and the turnaround team expertise ensures increased effectiveness and better practice
- *Drawbacks*
  - Planning is more complex and costlier
  - Integrated product to product network required
  - If network isn't adhered to, work may be delayed
  - More complex phased start dates for manpower
  - Greater risk of accidents/incidents than in other options

4. Total event

- *Features*
  - Turnaround manager is responsible for the total event, product to product
  - Turnaround team has its own operations and process personnel
  - The plant team hand the live plant over to the turnaround team who shut it down, overhaul it, start it up and hand it back to the plant team
- *Benefits*
  - Greatly reduced interface problems
  - Product to product duration can be significantly shorter and production loss less than other Options
  - Well practised team in total control can greatly increase effectiveness even over Option 3
- *Drawbacks*
  - Planning more complex and costlier
  - Product to product plan required
  - Lack of local knowledge may be a problem
  - Overrun costs borne by turnaround team
  - Risk of accidents/incidents greater than Options 1 and 2 but less than Option 3

Figure 6.13 Characteristics of the four SDSU options

The option chosen will depend upon the culture of the client's company and the constraints operating at the time, which (among other things) could include:

- plant team's expertise (How well would they know their plant?);
- Turnaround team's expertise (Especially the manager's);
- amount of money available (To buy in expertise);
- amount of time available (Would it permit a particular option to be implemented?);
- plant profitability (If high, would necessitate a quick Turnaround);
- plant utilization (May be low, so a lengthy low-cost Turnaround may be acceptable);
- plant complexity (May force the choice of a particular option).

The constraints have to be balanced and the options carefully considered. It cannot be stressed enough that this is one of the most crucial decisions that will be made during the preparation phase. It is worth repeating that it will have a fundamental effect on the planning and execution of the event. For that reason, the decision on which option to adopt must be taken by the steering group.

Having described the options available for combining the shutdown and start-up logic with the mechanical duration, it becomes possible to define the shutdown and start-up networks.

The shutdown network

The shutdown network (or plan) is a critical path programme and its supporting documents, which define (in logical sequence) the activities required to bring the plant off line and prepare it for handover to the Turnaround team.

The network is created by the plant team and should comprise the following elements:

- the shutdown logic for each plant system;
- identity of significant equipment within each system;
- the order in which each system will be shut down;
- the individual activities performed in the control room at each stage;
- the individual activities performed on the plant at each stage;
- whether activities are concurrent or sequential and how they are integrated;
- the duration of each activity;
- waiting time (if any) between specific activities;
- total duration for shutting down each system;
- total duration for the complete shutdown.

Other information required includes:

- temperature, pressure and flow-rate of fluids to be used;
- specification of any chemicals to be used;
- equipment, services and utilities required;

- numbers and types of manpower resources required;
- safety plan for the shutdown.

As with the planning of major tasks, shutdown networks must take account of every known aspect of the shutdown. If it is intended that mechanical work will be permitted on some systems while others are still in the process of shutting down then the earliest start times for mechanical work should be indicated.

Start-up network

The start-up network is similar in format to the shutdown network and all of the elements listed above for the shutdown are also specified in the start-up process. An additional requirement during start-up will be the testing programme which could include instrumentation tests, pressure tests, relief valve tests, product quality tests etc. – all of which must be written into the logic of the network.

Work scheduling

The planning sequence

In order to plan effectively we need to understand the sequence of events that take place between shutting the plant down and starting it up again and ensure that the plans and schedules match the actual sequence.

Figure 6.14 is an example of a templated planning sequence for a vessel on a chemical plant. It lists first the outline stages of the sequence (above the

Figure 6.14 Template of planning sequence (example – vessel on chemical plant)

line) and then breaks these down into smaller activities (below the line) in the manner of the work breakdown structure.

Each step is linked to the preceding step and the succeeding step in a continuous sequence so that when we allocate each activity a time and add those times together we get the duration for that particular task. There may be some overlap between some of the activities and that would be taken into consideration.

From here we can go two ways: we can either break each of the activities down into its natural steps to create a job method or combine the whole task with other tasks to create a schedule for the Turnaround.

These days, most companies create their Turnaround schedules on computer. There are a number of software packages available. The two main packages used in the UK are Microsoft Project and Primavera P3. As has been previously pointed out, care must be taken not to be seduced by the power and accuracy of the software into believing that all we need to do is input the data and the computer will do the rest.

One of the problems of using a computer is that it has generally reduced our understanding of what critical paths are and how they are put together. It is often left to the software to do this and sometimes it doesn't do it very well, despite what the manufacturers claim.

We can only minimize the duration of a Turnaround if we understand how the critical path that determines the duration was put together; understanding how each of the activities is linked to each other and being able to see it. Reliance on computers denies us this insight.

It is not within the scope of this book to discuss or compare the relative merits of different software packages; this information is readily available on the internet.

Figure 6.15 shows part of an event schedule with the basic information that would be provided by the computer. Each activity is coded, clocked and timed and a graphic representation is provided showing the connection between the activities. In this example, the critical path runs through code numbers S100–S135, S173–S185.

The network programme (see Figure 6.16)

When the logic has been worked out for shutdown and start-up, a bar chart is constructed showing the total shutdown and total start-up sequences and durations. The mechanical durations (especially those of the network tasks) are inserted between the shutdown and start-up elements so that a 'product to product' duration can be calculated, by summing lag time, shutdown duration, mechanical duration and start-up duration. The network task is the task which is completed last, and is not necessarily the longest duration mechanical task (an important point which is sometimes missed). The 19.75-day job which cannot be started until Day 4 will be the critical path, not the 22-day job which can be started on Day 1.

Code	Activity	Clock		Schedule 1	2	3	4	5	6	7	8	9	10
S100	Install battery limit blinds	1-0800	8										
S105	Dump inventory and blow down	1-1430	6.5										
S110	Steam out distillation column	2-1430	24										
S115	Open 3 column man-ways	2-1830	4										
S120	Allow column to cool and test atmosphere	3-0430	10										
S121	Install vessel isolations	3-0530	1										
S125	Install barrier scaffold above bottom trays	3-1330	8										
S130	Remove top 11 trays	4-2230	33										
S135	Install 11 top trays	6-1830	44										
S140	Remove fallen trays from dished end	3-1930	6										
S145	Install barrier scaffold above dished end	4-0330	8										
S150	Install bottom six trays	6-0330	48										
S151	Install platform scaffold on dished end	4-0930	6										
S155	Inspect weld under dished end	4-1130	2										
S160	Carry out weld repairs	4-2330	12										
S165	Radiograph welds	5-0330	4										
S170	Remove bottom barrier scaffold	5-0730	4										
S172	Remove platform scaffold	5-1130	4										
S173	Remove top barrier scaffold	6-2230	4										
S174	Carry out final vessel inspection	7-0330	5										
S175	Box up vessel man-ways	7-0730	4										
S180	Remove vessel isolations	7-1130	4										
S185	Start-up sequence	9-0130	38										

Figure 6.15 Part of an event schedule

Options

Depending upon the size of the event, the technology available, and the company's resources of skill, time, money and equipment, a work schedule can be constructed in several different ways, viz:

1. Using a software scheduling package on the client's mainframe computer.
2. Using a software scheduling package on a stand-alone PC.
3. Manually, using a project planning sheet.
4. Manually, using a 'shuffle board' (a slotted board – similar to those used to hold 'clocking in' cards – which lays out the different elements of task bars, using different coloured cards, to form a complete work schedule. Its advantage over the project planning sheet is that, when circumstances or priorities change, the elements can be 'shuffled' backwards or forwards to quickly represent the new situation.

Objectives

The objective of scheduling is to produce an integrated work programme (covering a period of anywhere between one and eight weeks) which will provide:

1. A plan for executing all tasks in a logical sequence.
2. A cost profile which will not exceed the budgetary profile.

JOB 'A'

JOB 'B'

	01	02	03	04	05	06	07	08	09	10	11	12	13	14	15	16	17	18	19	20	21	22	23	24	25	26	27	28	29	30

1. Job A has a mechanical duration of 22 days, Job B mechanical duration is 19.75 days and based on this alone, Job A would be considered the network job.

2. If the shutdown and start-up time for each job was known Job A would be 26.2 days long and Job B would be 25 days long – Job A would still be considered the network job.

3. However, it is only when the logical relationship between the two jobs is known that the true network can be calculated.

4. Due to the lag time at the beginning (Job B cannot be shut down until Job A is off line) and at the end of the mechanical duration Job B is the longer job by 1.5 days and is therefore the network.

5. The network is calculated from the first activity in the shutdown until the last item of equipment is back on line, and that is not necessarily the 'longest' job.

LAG TIME   SHUTDOWN AND START-UP DURATION   MECHANICAL DURATION

Figure 6.16 Networks

3. A duration within the available time scale.
4. A manpower-needs profile which can be resourced.
5. Economical and sustainable work patterns.

The final Turnaround schedule will be an optimized blend of the above requirements. The inputs required to build the schedule and the constraints within which it must operate are defined in Figure 6.17.

The constraints

The art of scheduling lies in the balancing of constraints. Consider this: if time, money and resources were unlimited there would be no need to plan or schedule work. It is constraints which define the planner's role.

There are two useful observations to make about constraints. The first is that they are intimately connected to one another. Take four of the basic constraints surrounding a Turnaround, viz duration, cost, resources and working patterns. Each has its own individual impact on the work scope and is, in turn, influenced by the work scope, but each also affects the others. If one changes, the others will either change automatically or will have to be changed to compensate. For example, one way of halving the duration of a Turnaround which had been planned on single-shift working would be to change to double-shift working – but that would also mean that twice as many people would be needed.

Figure 6.17 Work scheduling

The second observation is that 'constraints' constrain objectives, and objectives are often in conflict. One objective may be to complete the event in ten days. Another may be to use only single-shift working patterns to minimize the number of resources used and keep costs down, but the duration of the longest task in the work scope is planned at twelve days. Clearly, given the current situation, both objectives cannot be met. The constraints have to be balanced. This is the task facing the Turnaround manager and the steering group. It is their job to optimize the Turnaround plan.

Optimizing the Turnaround plan

Presenting the plan

Once all elements of the plan have been finalized by the Turnaround team, the Turnaround manager presents them to the steering group for final discussion and, it is hoped, for approval. Should the group reject the plan at this stage their reasons for doing so might well include one or more of the following:

- the duration is longer than they desire it to be;
- the cost is too high;
- the resources are greater than the plant saturation level;
- the working patterns are undesirable.

Optimizing the plan

In this case, the steering group need to balance the constraints and objectives. There are ten main options. The first eight are:

1. *Increase the duration of the event* – so that the amount of work required can be accomplished with less people or a less intensive shift pattern.
2. *Reduce the duration* – if it is thought to be too long. This will mean either increasing resources or altering the shift pattern. However, if the duration of the event is determined by that of the critical path task, it will not be possible to do so – unless the critical path task is eliminated.
3. *Increase the cost* – in order to reduce one of the other constraints. Again, this is not always possible.
4. *Reduce the cost* – this will mean reducing something else also.
5. *Increase manpower* – in order to reduce duration but, depending upon available resources, this may not be possible.
6. *Reduce manpower* – to reduce costs.
7. *Increase the use of shift work and overtime* – to reduce duration. Once again, this may not be possible.
8. *Reduce shift work and overtime* – to reduce costs.

These are conflicting options. The steering group may exercise any combination of them to achieve an 'optimized' plan. How they are exercised will depend largely on which constraints and objectives the group perceive to be the critical ones. Time? Money? Resources?

The ninth option is to reduce the work scope. Remember this is at the point where all tasks on the work list have been technically justified. If this option is chosen it is for business reasons not technical ones. In this sense, technically, it could be said that the steering group are taking a risk. They will have to bear the consequences if the work omitted results in an adverse effect on plant reliability at a later date.

The tenth option is more involved, much more challenging, but potentially very rewarding. It involves the challenging of current assumptions about how work is done. It is used primarily in the attempt to reduce duration but obviously has an effect on all other constraints. The Turnaround manager forms a team of those people he considers to be best suited to the task at hand. Their purpose is to challenge, using creative thinking techniques ('brainstorming'), every element of the longest duration jobs on the work list in order to reduce the time scale.

A long duration job consists of a sequence of different activities carried out by different people. The activities have been planned in a particular sequence, using specific methodology, technology and resources. Most of all, the plan is the visible result of a sequence of decisions made on the assumptions of an individual or a small group. The methodology or the technology, for instance, may have been chosen purely because 'that's the way it has always been done'. The creative session is organized to break through those types of assumption to a newer, more effective way of doing things.

What has to be asked is whether time can be saved by:

- eliminating the task altogether (alternatives? consequences?);
- altering the content of the task;
- altering the sequence of activities;
- performing sequential activities in parallel;
- reducing the duration of, or eliminating, any individual activity;
- reducing or eliminating gaps between critical path activities;
- altering either the number or the mix of manpower;
- altering the shift pattern on this particular task;
- using a different methodology;
- using a different technology.

Reality teaches that there are many ways of achieving the same end, only assumptions limit us to one way.

Contingency planning

The challenge

As discussed in chapter one, Turnarounds deal with plant that may be old, worn out, damaged or malfunctioning. For this reason, contingency planning is probably the most vexed and emotional element in Turnaround planning. So much so that it is often ignored – and the omission later regretted. The reason for this is that, in the context of a Turnaround, contingency planning is an activity which builds extra time, money and resources into a plan to cover for emergent work – tempered by the knowledge that it might never happen. This uncertainty leads to much debate about the need, and in extreme cases the ethics, of building preparation for contingency into a plan. On one hand if the Turnaround manager builds in what is considered to be too much contingency allowance he may be accused of 'padding out' or 'building fat into' the plan for his own comfort. He will get little support from the steering group especially if, in the event, the emergent work is less than predicted. On the other hand, if the Turnaround manager builds in no contingency, or too little, and a significant amount of such work does emerge, causing an over-run on time and costs, he will carry the responsibility for the business consequences of the over-run. As an example, take the task request for a particular vessel, which reads 'Inspect and if necessary repair . . .'. How should this be treated? The person who wrote the request has raised the possibility that something may be damaged or worn, and may need repair (which may be a significant one) but, when questioned, he states that he does not know for certain whether a repair will be needed but believes that it might. He has established his concern and it is now up to the Turnaround manager to act.

Does the Turnaround manager build in time and resources for such contingencies – and, if so, how much? He can consult with technical experts but they can only advise on how long a certain repair might take (all other things being equal) or how much it might cost (providing everything goes to plan). They cannot answer the two vital question he needs answered. Namely, will a repair be required? And if so, what will be the extent of it? They just don't know – nobody does – nobody can. The item in question is hidden from view. The Turnaround manager must therefore guess – nothing more scientific than that. The guess will be based on his (and other's) experience of similar past events, so-called 'engineering judgement' – but still a guess!

The one sure thing is that there will be an amount of emergent work. An informal survey of twenty Turnarounds performed on older plants showed a low of 5 per cent and a high of 47 per cent cost over-run due to emergent work. These figures are indicative only but a spread as wide as this, with a mean of 26 per cent, could lead to the 'too much – too little' situation described above.

In one actual case a large vessel had a suspected crack in the shell. A contingency allowance of two days was built into the plan for weld repair, heat treatment and radiography – it was agreed that this was a tight target – no one new just how tight. In the event, the crack was found but it was almost three metres long and fifteen millimetres deep. The repair was completed twelve days later.

A second case involved a packed column in which it was suspected that there was something wrong with the table that supported the packing. On inspection it was found that the table had capsized and was blocking the hole from which the packing was normally extracted to allow access to the table. One day's contingency allowance had been made for any necessary repair. The normal one-day activity of removing the packing took five days. The table was found to be undamaged but the time had already been lost.

As far as planning for contingencies goes, The Turnaround manager can only exercise his best judgement and get the commitment of the steering group to support the allowance calculated or take joint responsibility for altering it. Another practice is to build in a general contingency reserve for the whole event, but this too is only a guess and prone to all of the vagaries explained above. As a general rule, if contingency for emergent work is used, the money should be held in a separate fund, only used if required and any residue refunded to the business.

Defining emergent work

There are two types of emergent work: extra and additional.

Extra work – is generated by an existing task, as in the above examples of inspecting an item and finding a fault which must be rectified. It can also come about as a result of damage inflicted on the item by the team which is performing the task or by someone else.

Additional work – is tasks added to the work list during the event. This may happen because the plant team simply forgot to include them on the approved work list (even after all those work list meetings!), or because of something that happens during the event (this is most likely to occur when the plant is being brought back on line).

One common fault is that pumps – which may not have been included in the Turnaround – fail on start-up and must be replaced or repaired. Unfortunately at that late date there is no opportunity to recover the time lost.

Both types of work must be closely controlled (see Chapter 13).

Finalizing the plan

Several iterations of the process of optimizing the plan are likely to be needed before it is finally approved by the steering group. It should then be widely published so that everyone concerned with the event is able to peruse it and, if necessary, challenge any of its elements. Once key personnel have agreed to the plan, all subsequent activities must conform to it. The only exception being that if the situation changes, or an element of the plan is found to be ineffective, the steering group will reconvene to suitably adjust the plan. Using the approved plan the Turnaround team then generate a number of documents to allow them to monitor and control the flow of work during the event.

Generating control documents

The purpose of generating control documents is to:

- Quantify available resources. (How much? What skills?)
- Define work to be done on a daily basis.
- Track performance and work achieved.
- Flag up if the event is ahead of, on or behind target.
- Indicate future progress.
- Indicate need for corrective action.
- Maximize use of vital resources.
- Minimize expenditure.
- Indicate priority areas.
- Indicate the impact of unpredicted problems.
- Help the Turnaround manager to stay sane.

Typical documents generated are:

Reports	– Display current achievements, e.g. 'S' curves, bulk-work control sheets and schedule updates;
Worksheets	– Define workload, e.g. daily bar charts and punch lists;
Work control sheets	– The visible aspect of single point responsibility; Each step in a task is 'signed off' by the person responsible for it. The next step is not started until the signature is in place.

Having put together the Turnaround plan, the Turnaround manager needs to build an organization to execute it, a task which will be dealt with in the next chapter.

Case study

This case study will highlight that it is human beings who plan Turnarounds, not computers. While this may seem obvious to some, to others it may come as a surprise.

A particular Turnaround had been cancelled by the steering group, on the mistaken assumption that this would have no adverse effects on related plants. Six weeks before the programmed due date it was discovered that the Turnaround would indeed have to be done, and on the original date because it interacted with the operation of another larger plant.

A steering group meeting was hastily convened and the Turnaround manager was asked to put together a plan for the work (some 270 tasks, six of them major ones). There was no time to plan the event on computer, generate a schedule or produce the normal task sheets. A scratch team was assembled and – after the initial shock and denial stage was overcome – they agreed to put aside their previous training (which placed the computer at the heart of the planning and scheduling effort) and plan the event manually. The schedule was created by cutting, pasting and colour-coding sections of the work list (Monday's work coded yellow, Tuesday's green, and so on). This exercise also generated the critical path and duration. Major task instructions were written down on A4 sheets and the supervisors who would do the jobs closely briefed by the Turnaround manager. The remainder of the work was put on punch-list control sheets and the only detailed work which was carried out was the specification of materials.

The event was completed within duration and cost. At the post-event de-briefing, members of the team stated that, although it had been extremely hard work, they had felt more in control of the event because (a) they had personally contributed more to the logic of the plan during the preparation phase, and (b) the fact that they did not have a computerized schedule to update daily meant that they spent a lot more time on-site, monitoring, advising and assisting.

Admittedly this was a relatively small event but it did bring home the point that the intelligence in planning and scheduling resides in the human beings and not in the computers. Our increasing dependence on computer systems – hugely effective as they are – should not blind us to this fact.

7

The Turnaround organization

Introduction

The Turnaround organization – the blend of people who will execute the work on the day – is critical to the success of the event. Management must think thoroughly through the issues involved, select the most suitable people available at the time, and organize them with great care in order to forge the strongest possible organization for controlling the Turnaround. The organization is the template which will shape the course of the Turnaround and any flaw or weakness in it will infect every other aspect of the event.

Organizational combinations (see Figure 7.1)

The shape and size of the organization will be determined by separately addressing three key questions, i.e.:

1. Who will manage the Turnaround?
2. Who will carry out the work?
3. What will be the relationship between the two?

Taken together (as they must be), the answers to these questions will reveal that a large variety of organizations is possible.

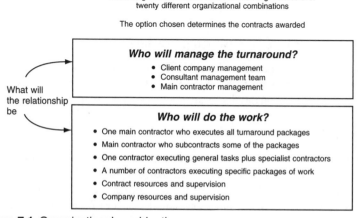

Figure 7.1 Organizational combinations

Who will manage the Turnaround?

As explained in Chapter 2, there are three options available when choosing a Turnaround manager, the choice being based on a number of considerations.

Option M1 (where M = Manager)

Select one of the company's own management or engineering staff. The considerations here will be ones of availability and competence. A kind of half-way house between this option and M2 is to select a member of staff to manage the Turnaround and then bring in an experienced consultant to 'advise' him.

Option M2

Bring in a consultant management team. More and more companies offer this service. The considerations are price, (they can be expensive) and company culture (do we really want to put our future into the hands of strangers?).

Option M3

Bring in a main contractor to manage and execute the bulk of the work. The consideration here is one of trust. Whereas the consultant management team may be strangers they are at least independent of the contractors who will execute the work. On the other hand the contractor manager has a vested interest which might lead him to maximize profit or take decisions which are favourable to his company and may be to the disadvantage of the client. Again, company culture will determine if this is a viable option.

Once the manager (or management team) is chosen, the next question is:

Who will do the work?

There are six main options to consider;

Option W1 (Where W = work organization)

Employ one main contractor who will execute all work packages. World-wide, several companies claim to be able to handle the total package of work. At present (2005), their quality is variable but has generally improved since 1999.

Option W2

Employ one main contractor who employs, and manages, sub-contractors for a number of packages. This has the advantage that the Turnaround manager oversees only the interface between the control team and the contractor. The main disadvantage is that it introduces another layer of management and can complicate communications.

Option W3

Employ one main contractor, executing general tasks, plus several specialist contractors. This is similar to Option W2, but in this case the Turnaround manager interfaces directly with the main contractor and the specialist contractors. The advantage is a flatter organization. The disadvantage is that the Turnaround manager has many more people to deal with.

Option W4

Employ several contractors executing specific packages of work. Again this is similar to option W3, but there is no main contractor doing the bulk of the general work. The advantage is that the Turnaround manager has an even tighter interface with each contractor. The disadvantage is that because a main contractor would normally handle the semi-skilled and unskilled work within his contract, his absence means even more contracting companies will need to be used, further increasing the number of people the Turnaround manager has to deal with.

Option W5

Rely on contract agencies only to provide manpower, the company supervising it, and all other resources, itself. From a client's perspective the advantages are that he directly controls the labour and therefore the execution of the work, and does not have to pay all the contractor management fees. The main disadvantages of using agency labour were discussed in Chapter 5 ('Using contractors – the downside'). There is a further one, however; supervising agency labour can be very difficult, especially when undertaken by plant-based supervisors who are used to leading a small efficient team that (to a great degree) co-operates with the supervisor. Contrast this with the large amount of agency labour needed to perform a Turnaround, and consider also the following observations:

- Agency workers will be strangers to the plant, and are only interested in the Turnaround as a temporary source of income.
- Productivity can be very low, which usually triggers the emotional reaction to 'throw more men at the job', thus complicating matters even further.

Option W6

Use one's own resources and supervision. This is not normally an option for the total package of work. Most companies will have their own staff performing some of the tasks – almost certainly higher voltage electrical work. The advantage in using local knowledge must be weighed against the disadvantages of labour agreements between management and unions regarding shift patterns, overtime and special payments (probably ideal for working on an operating plant) which can be a hindrance to the effective performance of the Turnaround.

Clearly, the number of different possible organizational structures that can be created by combining any of the 'M' options with any of the 'W' options is not small (see Figure 7.1)

What will be the relationship between the two?

Because of the many combinations possible it is important to consider relationships, as these will influence performance.

If, for example, a client manager is appointed to manage a number of contractors does he keep them segregated or form them into one team?

Basic principles

The steering group and Turnaround manager are responsible for the effective execution of the Turnaround. They must ensure that all of the elements that go together to make up the event are properly organized. A number of principles have been developed out of past experience to guide them in their endeavours.

A Turnaround is a task-oriented event

The work list, a series of tasks, is the foundation of the Turnaround and everything else depends upon it. For this reason, the task is put at the centre of the planning and preparation effort. All other elements are directed towards the safe, timely and cost-effective performance of the task. This approach generates the particular methodology put forward in this book.

The minimum number of people should be used

A Turnaround (especially a major event) involves a large number of people. Careful analysis ensures that only those people who are necessary are employed on the event. Problems are not solved by 'throwing bodies at them' and it is more preferable to have a smaller number of higher skilled personnel than vice versa.

The principle is based on the following criteria:

- Each plant has a saturation point beyond which the employment of more people will serve only to decrease useful work done rather than increase it.
- Increasing the number of people increases the risk of communication difficulties, and of potential hazards, both to personnel and to their environment.
- The number of highly skilled people is limited, so increasing the number of people on site risks decreases in the overall skill level.
- Costs are increased in proportion to the increase in people.

A Turnaround organization is hierarchic
(see Figures 7.2 and 7.3)

Much has been written in the last twenty years about the limitations of the traditional bureaucratic management structure and the virtues of matrix management and open teams. However, due to the hazard and complexity of Turnarounds the organizational model that has, so far, worked best is the hierarchic pyramid. It is effective so long as the responsibilities shown in Figure 7.2 are properly discharged and there are effective communication links between levels. The number of levels in the hierarchy will depend upon how large and complex the event is.

Figure 7.3 highlights some of the issues surrounding the selection of key members of the organization. The comments in the various boxes are based on the author's experience and do not purport to be universal. In different situations, in other companies and, indeed, in other countries, different imperatives may apply to the selection of the organization.

Figures 7.4 and 7.5 show models of Turnaround organizations. Figure 7.4 models a large event in which the plant was split geographically and the organization was a mixture of plant-based, contract and project personnel. Because it was a large event with a high technical content, it required a large management and technical team (this is what makes performing work on a Turnaround more expensive than performing it during normal operations). Note the requirement to have one co-ordinator spanning two areas due to the shortage of competent personnel.

Figure 7.2 Execution organization hierarchy

Organizational level	Parent company	Consultant company	Plant personnel	Main contractor	Sub contractors
Strategic level (Turnaround policy team) – provide funds – balance constraints – set objectives – formulate policy – delegate authority – monitor and change policy	Senior engineering, business and financial managers bring their specific expertise and understanding of constraints to the team in order to set smart objectives and create policies to achieve them.	Turnaround manager brings technical expertise in work definition, planning and the management of big events. Acts as the agent of the turnaround steering group.	Plant management provide local knowledge of plant requirements and problems. Responsible for providing the work lists.	Contractor management are only invited to attend steering group meetings in special circumstances and then only to present information and answer questions.	No involvement.
Project level (Turnaround manager) – translate policy – formulate plan – manage the event – delegate activities – monitor and report – change policy	On very large or complex events the steering group may appoint a senior manager to take overall responsibility. The plant and turnaround managers run the event and report to the senior manager.	The consultant turnaround manager possesses skills, specialist knowledge and experience obtained from full time involvement in turnaround management.	The steering group or plant manager may appoint a manager or engineer from the plant to manage the turnaround. Can also be the maintenance manager.	The main contractor may offer the option of a total package including the full management and planning team. This should only be considered if the company has a proven track record.	No involvement.
Sub-project level (Area engineer) – manage activities – provide expertise – delegate tasks – monitor and report – solve problems	No direct involvement but may be able to second people from other plants to perform this role.	Consultant companies, when requested, will provide very experienced engineers in key positions to manage areas of turnaround. This strategy is to enhance local performance and train local engineers.	Subject to availability, the plant manager may offer the services of plant engineers to perform this role. Not normally experienced on large events.	The main contractor will normally offer the services of experienced engineers who will work under the turnaround manager.	Specialist sub-contractors may provide engineering coverage for large projects. It is not normal for general sub-contractors to offer the service. This option should be used only as a last resort.
Task level (Area co-ordinator) – co-ordinate tasks – control activities – delegate sub-tasks – monitor and report	No direct involvement but may be able to second people from other plants to perform this role.	Will normally provide the planning co-ordinator/officer. Some consultant companies will offer the services of experienced on-site, area co-ordinators.	Subject to availability and need, the plant manager may offer a senior planner as planning co-ordinator. Other trade co-ordinators subject to availability.	The main contractor will normally offer the services of co-ordinators.	Sub-contractor normally offers the services of a general foreman who will act as a co-ordinator. This option should be used only as a last resort.
Sub-task level (Supervisors) – allocate work – supervise work teams – troubleshoot – monitor and report	No direct involvement but may be able to second people from other plants to perform this role.	No direct involvement but can usually recommend manpower agencies which can provide people.	Electrical and instrument supervisors are normally provided by the plant manager. Supervisors for other disciplines are subject to availability and need.	The main contractor will provide supervisors.	Sub-contractor provides supervisors or responds to contractor supervisors.
Work level (Tradesmen and semiskilled) – perform work – report problems	No direct involvement but may be able to second people from other plants to perform this role.	No direct involvement but can usually recommend manpower agencies which can provide people.	Subject to availability, the plant manager may offer maintenance personnel to perform this role.	The main contractor may provide manpower or hire in sub-contractors.	Sub-contractor provides his own manpower.

Figure 7.3 Organizational issues to be considered

Figure 7.4 A sample Turnaround organization

Figure 7.5 Example of an organization for a small Turnaround

Figure 7.5 models the organization for a much smaller event (consisting mainly of the overhaul of a large number of small pumps and valves) which was manned by plant-based personnel supplemented by some agency semiskilled and unskilled labour. This is typically the type of event chosen for a Turnaround manager's first performance in the role, to build up his experience. Although the organization is smaller, note that all of the elements are still represented, albeit in a simplified manner (the 'manager' also carries out the duties of the engineer).

One person must be in overall control

This is a natural consequence of the last criterion. A Turnaround is such a compressed event, with much more work being done at higher than normal expense by a large number of people in the shortest possible duration. As a result, there is little time for deliberation or discussion. Many decisions have to be made immediately. To make this possible, one person must be put in overall control of the event.

Single point responsibility is exercised at every stage
(see Figure 7.6)

Much of the work done on a Turnaround is of long duration and involves a number of people from different skill backgrounds, departments and even companies. The Turnaround, being a process, is defined as 'A sequence of activities performed in logical succession', which prompts the following observations:

- To ensure that the logic of the plan is followed, each activity must be carried out at the correct point in the sequence.
- Each activity must be executed correctly and, if necessary, inspected to verify it.
- There is a gap between each activity. It must be controlled to avoid wasting time.
- If consecutive activities are carried out by different individuals there must be an effective handover.

Throughout the life of the task the responsibility for ensuring that it is performed effectively may change many times – rather like passing the baton in a relay race, so there must be a mechanism to ensure that the baton is not dropped. This mechanism is called single point responsibility and it requires that the persons responsible for each activity must ensure that:

- A good handover is obtained from the person responsible for the previous activity.
- The activity that the person is responsible for is properly executed.
- The task is properly handed over to the person responsible for the next activity.
- The completion of each stage must be recorded.

The Turnaround manager is ultimately responsible for ensuring that all tasks are carried out properly. He must therefore ensure that the persons

Single point responsibility requires that the person in control of the task at any stage has sole responsibility for:

- accepting the task from the previous stage;
- completing the current stage to plan;
- handing the task over to the next stage;
- reporting stage completion.

The turnaround manager has sole responsibility for ensuring all stages are linked in an overall plan and everyone is briefed on it.

Work sequence for a small task, i.e.: a vessel with an internal crack

Stage	*Responsible person*
Validate task	Preparation engineer
Plan task	Turnaround planner
Validate plan	Preparations co-ordinator
Schedule task	Planning co-ordinator
Validate schedule	Turnaround manager
Obtain resources	Turnaround planner
Brief personnel	Turnaround manager
Shut down plant	Plant manager
Isolate vessel	Maintenance supervisor
Open vessel	Task team leader
Test atmosphere	Safety officer
Install lighting	Electrical supervisor
Install ventilation	Task team leader
Install scaffold	Scaffold supervisor
Inspect/discover defect	Inspector
Troubleshoot problem	Turnaround manager
Create repair procedure	Design engineer
Carry out repair	Welding supervisor
Inspect – NDT	Inspector
Pressure test	Maintenance supervisor
Accept repair	Inspector
Strip out ventilation	Task team leader
Strip out scaffolding	Scaffold supervisor
Strip out lighting	Electrical supervisor
Box up vessel	Task team leader
De-isolate	Maintenance supervisor
Sign completion form	Plant engineer
Trip and alarm test	Instrument supervisor
Start up plant	Plant manager

Figure 7.6 Single point responsibility

responsible for each activity are clearly identified, properly briefed, and provided with the means to receive and hand on their tasks.

Every task is controlled at every stage

To assist in the operation of single point responsibility, a system of 'sign off' control documents (see Chapter 13) is used on particular tasks. As each critical activity is completed a responsible person signs it off and dates it on a control sheet. The sheets are prominently displayed in the control cabins and are continuously monitored and updated. The display allows everyone to see the current state of the tasks involved and spot any task which has either fallen behind or has a 'hole' – a space where an activity has not yet been signed off.

The organization is a blend of the required knowledge and experience

The optimum organization would blend the following:

- *plant personnel*, who possess local knowledge;
- *Turnaround personnel*, skilled in planning, co-ordination and work management;
- *technical personnel*, who possess engineering, design and project skills;
- *contractors*, and others, who possess the skills and knowledge to execute work.

In a real situation, the optimum may not be achievable. The Turnaround manager, with the assistance of the policy team, must build the best organization possible with the available personnel. To this end, they must be able to assess the existing personnel, recognize its strengths and weakness, and take steps to maximize the strengths and minimize the weaknesses.

Constraints

There is no standard pattern, shape or size for a Turnaround organization. For a particular event these characteristics will depend upon the constraints operating at the time. There are many such constraints, all of which have a greater or lesser influence on the form of the organization created. Typical constraints include, but are not limited to:

- the amount of money available to plan and execute the work;
- the size and complexity of the Turnaround;
- the timing and duration of the event;
- the availability of personnel;
- the company culture and norms.

The task of the Turnaround manager, working within current constraints, is to consider the options available and to design an organization which, in his estimation, best meets the needs of the event. He must then propose this organization to the steering group and inform them of any unavoidable weaknesses and their likely effects.

In the light of any exposed weaknesses, the task of the steering group is to test the proposed organization against their set objectives and to take one of four courses of action, viz:

- *approve the organization* as proposed, recognizing any weaknesses which may impair its capability to achieve the pre-set objectives;
- *remove the constraints* so that the organization is better suited to achieve the pre-set objectives;

- *change the objectives* to suit the current constraints and organization;
- *change the organization* to better suit the objectives and constraints.

This requires a realistic approach. In rare cases the steering group may demand the achievement of objectives which are not attainable but refuse to remove the constraints. The Turnaround manager should avoid being pulled into this type of fantasy situation because, in the cold light of a missed objective to which he has committed himself, he will be left to take the consequences, whatever they may be. In the final analysis, the success or otherwise of the organization depends crucially on the balance between constraints and objectives.

Scarce manpower resources

Some resources may be scarce. While it is true that, in some situations, they may all be 'scarce', what is referred to here are those – e.g. coded welders, versatile instrument technicians, machines fitters – which are scarce by reason of the type of work involved. They may also include specialist vendors such as installers and servicers of digital control systems. The work thus involved should be classified as 'critical tasks' and could include, among others

- overhaul of machinery (especially of older models);
- relief stream inspection and registration;
- inert entry (especially in vessels with complicated internals);
- high voltage electrical work (which normally requires a licence);
- handling material which may be radioactive, explosive, inflammable or toxic (and which may therefore require a licence);
- coded welding (especially where exotic materials or rarely used techniques are involved);
- calibration of instruments (and electronic and computerized equipment);
- trip and alarm testing (a balance of expertise and local knowledge);
- protective coating and specialist painting (which may look simple but isn't).

Task complexity, the need for technical knowledge, specialized trade qualification and legally required licensing, limits the number of individuals who can perform critical tasks or are allowed to. Such personnel have the following characteristics:

- they form the top few per cent of the 'normal distribution' for their skill;
- they are usually highly paid and in great demand;
- they are usually available for a limited period only;
- they are typically focused on a very narrow band of tasks (often on only one).

Because the above personnel are defined as critical, they must be managed with care. Creating a critical manpower resource profile will help. A sample routine for this task would be as follows.

1. For each separate area of the Turnaround make out a list of the tasks which will require scarce resources.
2. Using best judgement (remember this is done at an early stage) assign a programme day when the resource will be required (e.g. Day 4).
3. Assign a programme time for the use of the resource (the best that can be done here is probably allocation to the morning, afternoon, evening or night shift).
4. Define the equipment required to perform the task (e.g. TIG weld set).
5. Define any logistic requirements (e.g. a generator, or rod ovens).
6. Define any dependent activities (e.g. radiography, pressure testing).
7. Identify any potential contingency (e.g. a repair or re-weld may be required).

Once the above list has been created the data is organized to create the profile as follows:

1. Create a chart with programme days along the horizontal axis and required manpower resource numbers on the vertical axis.
2. Record the resource requirements for each programme day.
3. Note the requirement for similar work on different tasks and areas for each day.
4. Manipulate the times so that the maximum amount of work is done by the minimum manpower resource.
5. Consider changing the timing of certain jobs (within the confines of the overall schedule) to move manpower from one day to another.

This exercise is initiated in the early days of the preparation phase to give an indication of the likely manpower requirements on critical tasks. The first run through will be pretty crude but, as better planning and scheduling information becomes available, the profile can be made more accurate. The manager should exercise judgement on these issues because the scarce resources may have to be procured before there is enough accurate data available – that is their nature!

Once the scarce manpower resources have been identified, the following facts should be established:

- the various providers of such resources;
- numbers likely to be available from each source during the event;
- any factors that will affect resource availabilities;
- the lead time required to hire each resource;
- alternative providers if the primary source dries up;
- the lead times for increasing or replacing resources;
- differentials in pay scales for manpower resources from different suppliers (ignorance of this has caused industrial action during an event because people were not happy about doing the same job as someone else, but for less pay).

Scarce resources may be obtained:

- from within the client company, either on the plant in question or at another location;
- by hiring from another company in the same business as the client (a competitor?);
- from equipment vendors (often overseas and very expensive);
- from a specialist sub-contractor;
- from the main contractor;
- in-house, by providing specific training for the work.

Even if the manpower resources are apparently available, the following measures will minimize uncertainty:

- arrange for the scarce resources at the earliest possible date;
- check all paper qualifications;
- if possible, test competence in the specific task required (welders have to do this all the time!);
- monitor and record performance and retain for future reference.

The lack of one scarce resource on the network task can extend the duration of the Turnaround, and what may seem to be an issue involving only a couple of people and a relatively small amount of money can have an expensive consequence involving those people who have to be retained during the over-run (and, often much more expensive, lost production).

Case study

The organization shown in Figure 7.4 was created to suit the needs of a Turnaround. Its form was dictated by management's response to the constraints that operated at the time. The main features were as follows:

A consultant Turnaround manager was hired to manage the mechanical duration of the event.

(*Constraint*: the client had no one with the requisite knowledge and experience.)

The plant manager controlled the shutdown and start-up phases.

(*Constraint*: in line with the company's safety first culture the senior manager decreed that the entire plant must be off line before any mechanical work could begin. The plant manager was the only person with the experience needed to manage these phases.)

A plant-based quality team was created to check and sign off the breaking and re-making of joints.

(*Constraint*: in the past, there had been a significant problem with leaking joints on start-up, so this additional team was imposed upon the Turnaround manager.)

A large up-grading project, carried out by contractors, was controlled by the company's project department.

(*Constraint*: the senior manager hired the Turnaround manager to directly manage only the plant maintenance overhaul and considered the project to be plant improvement.)

The plant was heavily instrumented so the control electrical work was treated as a separate 'area'.

(*Constraint*: the senior manager considered control electrical work to be of a critical skill level that was much higher than that of the mechanical work.)

The company had hired a contractor to handle one of the areas.

(*Constraint*: the company did not have enough internal staff to handle the event.)

The Area 1 co-ordinator also co-ordinated Area 2.

(*Constraint*: shortage of experienced co-ordinators.)

Planning and updating was centralized rather than being done by area planners.

(*Constraint*: insufficient funds to hire the extra staff needed for the area planning teams.)

The main effects on the event of the organization were that the Turnaround manager, the plant manager and the project manager did not see eye to eye on a number of issues, most importantly the sequencing of work in the few days leading up to the plant start-up. Because none of them was in overall control, their disputes had to be arbitrated by the senior manager, who was not technically trained. He tended to back the plant manager because he was the only 'company man' (the other two having been hired in) and this situation led to a duration over-run.

The quality of work in Areas 1 and 2 suffered because the co-ordinator – who was supposed to deal with both areas – was over-worked and unable to co-ordinate either area effectively. The number of leaks did fall drastically due to the close monitoring of the quality team, and the instrument work – especially the trip and alarm testing during start-up – was very successful.

Could the organization have been designed differently to make it more effective? The organization which was created to control the event was heavily influenced not only by the constraints operating at the time but also by senior management assumptions which coloured the decisions about who should do what. As can be seen from this case study, it is not so much the constraints which dictate the final shape of the organization so much as the management's responses to those constraints. Different responses would have created a different organization.

8

Logistics

Introduction

The business of logistics, in the context of a Turnaround, is reception, storage and protection, issue and demobilization of every item of material, equipment, services, accommodation facilities and utilities required for the event. Put simply, the logistic activity is aimed at ensuring that the right thing is in the right place at the right time and in a fit condition to perform its function. Logistics encompasses all of the non-technical elements of the Turnaround (the technical elements being defined as the activities required to carry out the tasks on the work list) even though it deals with technical equipment. It concerns the disposition of the many thousands of items, large and small, that are required by the technical teams to perform their activities in a timely manner. Logistics therefore affects everyone employed on the Turnaround.

Because of the above requirements, logistics must be planned and prepared with a rigour equal to that imposed on technical planning; this is crucial to the success of the Turnaround. Poor logistics can ruin the best devised technical plan and thereby prevent the achievement of the Turnaround objectives.

The logistics team

The logistics team is charged with planning and operating the logistics programme before, during and after the event – and is led by a logistics officer whose duties include, *inter alia*, the following:

- drawing up the master site plot plan;
- liaising with plant personnel and the Turnaround planning team;
- arranging all outside laydown areas;
- setting up stores and receipt/issue procedures;
- receiving, locating, maintaining and mobilization of:
  - materials proprietary items and consumables,
  - tools and equipment,
  - services and utilities,
  - accommodation and facilities;
- organizing site infrastructure;
- organizing the movement of bulkwork;
- the control of hazardous substances;
- providing for the daily needs of personnel.

The logistics co-ordinator reports directly, and refers any operational issues, to the Turnaround manager. The logistics team, normally made up of material marshals, storemen and drivers, assists the logistics co-ordinator in the execution of the logistics plan and reports directly to him.

The elements of logistics

There are two elements to be considered. The first is the actual physical object or substance to be dealt with and the second is its current disposition.

Item identification (see Figure 8.1)

This element refers to all of the items which are required not only to carry out the technical tasks but to take care of the day-to-day needs of the personnel employed. It includes, among other things, the following:

Materials	–	steel, wood, plastic
Proprietary items	–	valves, pumps, electric motors
Consumables	–	welding rods, filters, cartridges
Equipment	–	radiographic, torquing, bolt tensioning
Tools	–	electric and pneumatic hand tools, spanners
Transportation	–	personnel carriers, wagons, low loaders
Cranage	–	tower cranes, mobile cranes, hoists
Utilities	–	electricity, gas, water
Services	–	scaffolding, insulation, water washing
Accommodation	–	offices, stores, conference rooms
Facilities	–	toilets, changing rooms, mess rooms

It is vital that all of these physical items are acquired. It is equally as important that the current disposition of any item be controlled so that, at any given time, the questions listed in the next section can be answered.

The current disposition of the physical objects or substances (See Figure 8.2)

During a Turnaround, among the most commonly asked questions are the following:

Disposition stage	*Question to be answered*
Receiving on site	Has it arrived?
Accommodating or storing	Where is it?
Protecting and maintaining	What condition is it in?
Moving around site	Can I have it here, now?
Disposal	Is it off hire?
Recording of current disposition	Where is the proof?
Communicating with others	What is going on?

Typical lists of items and services acquired, accommodated, maintained and supplied by the logistics team include, but are not limited to:

Materials, proprietary items, consumables

- ferrous/non-ferrous metals, wood, plastic, paper
- fabrications, piping, flanges, flat or formed plate
- pumps, valves, motors, instruments and spare parts
- joints, nuts, bolts, studs, washers and pins
- electrical equipment, bulbs, wire and switches
- welding electrodes and wire, solder and flux
- bottled air, oxygen, acetylene, nitrogen and hydrogen
- de-ionized, demineralized and distilled water
- seals, filters and desiccants
- dye penetrant stain and developer
- paper overalls, gloves, overshoes and plastic bags
- lubricants, detergents and cleaning materials
- diesel oil and petroleum

Equipment and tools

- welding rectifiers, generators, compressors
- diesel oil, petrol and water tenders
- welding torches and oxyacetylene torches
- hoists, pulleys, pull lifts and hydraulic jacks
- cables, hoses, ropes, chains and slings
- ramps, barriers, zip-up scaffold and ladders
- bolt tensioning and torquing gear
- on site machining equipment
- inspection instruments and measuring equipment
- radiographic, ultrasound and MPI equipment
- mechanical, electrical and instrument tool kits
- specialist machines, jigs, fixtures and tools
- water washing hoses and lances
- skips, barrows, bogeys, brushes and shovels

Shut down/start up equipment

- special pumps, valves, bobbins tools and fixtures
- safety equipment and protective clothing
- hoses for washing and draining
- barrels and trays to catch liquid spillage
- boxes and bags to catch solid spillage
- detergent for cleaning oil or product spillage
- equipment for cleaning drains and culverts
- skips and trolleys for removing solids
- slip plates, blanking plates, nuts and bolts
- special plates for corrosive product
- temporary platforms, scaffold and ladders
- compressed air and instrument air
- adequate gas, electricity and mains water
- de-ionized, demineralized and distilled water

Cranage and transportation

- large semi-permanent cranes
- small mobile cranes
- hy-ab trucks
- heavy trucks
- flat bed trucks and tenders or trailers
- panel trucks
- small vans
- four wheel drive vehicles
- electric trolleys and fork lifts
- ambulances and fire trucks
- minibuses
- large buses
- motorcycles
- bicycles

Utilities and services

- mains electricity, gas and water
- water and chemical washing, blast cleaning
- inert entries, air moving, lighting, ventilation
- scaffolding, cladding and sheeting
- insulation (lagging and de-lagging)
- in situ machining and cutting
- air-arc, plasma and water jet cutting
- IG, MIG and plasma arc welding
- photogrammetry, laser aligning, acoustic ranging
- ultrasound measurement and thermal imaging
- introscopy and remote cameras
- radiography, ultrasound and MPI inspection
- bolt tensioning, torquing and testing
- ceramic metals and metal spraying
- personnel lifts, hoists and manveyors
- cleaning and scrap removal

Accommodation and facilities

- temporary offices, messing rooms, stores, locker rooms, washing and toilet facilities
- induction, first aid and safety cabins
- temporary pathways and duck boards
- electric supplies, wiring and connections
- water supplies, pipes and connections
- telephones (inc. lines), radios (including batteries)
- computers, photocopiers and facsimile machines
- office furniture, fittings and equipment
- messing facilities and provision of meals
- paper towels, soap and toilet rolls
- fire and toxic alarm facilities
- transportation of personnel to, and from work
- food and beverages
- cleaning services for accommodation

Figure 8.1 Physical objects and substances

The aim of the logistics co-ordinator and his team is to prevent the need for such questions (by knowing the numbers and types of physical items that are on site and the current disposition of each), and where that is not possible, to be able to answer them and other related questions with the minimum of delay.

The following issues must be considered

Receiving on site
- visually check quantities against purchase order
- query partial deliveries with procurer
- visually check for damage or deterioration
- arrange for technical inspection
- quarantine nonconforming items and arrange for concession or return to sender
- inform procurer of disposition
- cross check order, delivery note and invoice
- pass documents to procurer for processing
- record reception in ledger

Accommodate or store
Provide the following:

- adequate accommodation for people
- adequate storage space for small items
- lay down areas for large items
- locations for services such as water washing, scaffolding, insulation and other service equipment
- secure areas for hazardous substances
- locations for cranes and other heavy equipment
- locations for rectifiers, generators, compressors
- safe parking space for all vehicles
- rubbish skips, bins and bags

Protect and maintain
- provide fire and toxic refuges
- provide first aid posts and ambulances
- provide secure storage for valuable items
- protect items against corrosion and impact
- provide clean conditions for electronic gear
- check shelf life of perishable items
- service and lubricate mechanisms where required
- use correct storage procedures for gaskets, seals and other delicate items
- use good handling procedures to avoid damage
- repair or replace any damaged item
- provide protection for items stored in the open

Move items around site
- control the movement of all vehicles
- nominate approved routes for the movement of personnel, vehicles and equipment around the site
- provide safe routes for hazardous loads
- prohibit the use of hazardous routes
- prohibit permanent blocking of approved routes
- if routes are blocked, provide alternative routes
- procure a priority movement list at busy times
- avoid causing damage to items being moved
- lay on vehicles for overtime and shift work
- protect, maintain and regularly refuel vehicles
- protect pipes and cables which lie across roads

Dispose of hired equipment and services
As soon as possible because, during every extra hour on site, they:

- will cost money
- will take up space
- will be liable to be damaged
- will be liable to be misused
- could represent a hazard

Make arrangements with suppliers so that:

- your liability ceases the moment the equipment is off-hire
- supplier is responsible for removing it from site

Record current disposition
The logistics co-ordinator must know where every item is at any given time. There are thousands of items spread around many locations in various stages of use. Therefore, the logistics co-ordinator must have an effective procedure which shows:

- date received and present location
- items damaged, quarantined or returned to sender
- end user informed of arrival
- date issued, quantity and receiver
- remaining stocks and minimum stock levels
- hired items and services off-hire date

Communicate current disposition
The logistics co-ordinator must communicate effectively with other key personnel to inform them of:

- the arrival of items and services
- current condition (quarantined/damaged?)
- current location (stores/lay down area)
- current disposition (stored or in use)
- estimated time of arrival at on-site location
- unpredicted shortages (and strategy to overcome)
- damage or deterioration (and steps being taken)
- minimum stock levels (reorder of critical items)
- items and services off-hire (final cost)

Any other issues

Figure 8.2 Current disposition

In order to organize such a vast amount of information, the logistics co-ordinator requires some sort of conceptual model into which most, if not all, of the data required to effectively supply the necessary items for the Turnaround can be inserted. Supported by various schedules and procedures, the model used is the plot plan.

The plot plan

The plot plan is one of the most important documents of the Turnaround. It is, to the non-technical aspect of the event what the schedule is to the technical aspect. A good plot plan will afford the logistics co-ordinator the opportunity to control supply. A bad or non-existent logistics plan may lead to confusion at best and chaos at worst. This, in a business context, leads to over-run, overspend and a negative effect on quality. When the plan is completed it will display the location of every important element on the Turnaround.

Purpose

The purpose of the plot plan is to ensure the safety, availability and effective mobilization of every item on site. To achieve this it is necessary to organize the current disposition of every physical element involved (as already discussed in this chapter).

Preparing the plot plan

To prepare the plot plan, the logistics co-ordinator must survey the site and gather a large amount of information from plant and other personnel. There may exist a plot plan from a previous Turnaround and, if so, he may use it as a basis for his current plan. Failing this, there should be a basic plot plan of the plant, showing the position of all the critical elements. Some such plans are very detailed while others contain only a minimum of information.

Where a plot plan does exist, the logistics co-ordinator should not assume that it is accurate. Many plans are the original ones issued when the plant was commissioned (perhaps as much as twenty-five years or more in the past) and which have not been updated to reflect the many subsequent modifications. In the worst case there will be no plot plan and the logistics officer must create one.

It is therefore in the best interest of the logistics co-ordinator to survey the plant and gather information – to ensure that the basic Turnaround plot plan will be accurate. Finally, it must be drawn to a scale large enough for individual elements (located via a grid-reference system) to be identifiable.

The stages

There are three stages involved in producing a plot plan, viz:

1. drawing up the basic site map;
2. drawing in the Turnaround overlay;
3. issuing the plot plan.

Drawing the basic site map

Whether the basic site map is provided by the plant or created by the logistics co-ordinator, it must show such important features as:

The perimeter of the plant and the boundaries of the available land (if any) surrounding it – because a great deal of work will be carried out in a short space of time, the logistics co-ordinator must have a clear indication of what land is available to lay down the goods and services required.

All major items of plant equipment and connecting pipe work – in the correct configuration to ensure that available land adjacent to hazardous areas, areas of special function, or areas with difficult or restricted access, are identified.

All roads (in particular, public ones) which run through the site and/or the surrounding land – so that any interface with the public can be made safe.

All access to site and site roads – such as personnel, goods entry and emergency evacuation gates. During the event the number of people and the volume of goods and services entering and leaving site will be many times greater than normal and extra gates may have to be created.

The location of all fire assembly points, toxic refuges, fire fighting equipment, alarms, site telephones, emergency showers and eye baths – to ensure that such provision is adequate and that all Turnaround personnel can be shown the locations of this equipment as part of the site safety briefing.

Any other permanent or unusual feature – which could affect the logistics (e.g. heights of pipe bridges).

Once this basic map has been completed the logistics co-ordinator then has the task of applying all of the data which will indicate the disposition of the site, during the Turnaround, to the basic map. This is called the *Turnaround overlay*.

The Turnaround overlay (see Figure 8.3)

The logistics co-ordinator accurately marks the position, and codes the function (by colour, numbering, or other means), of every element added to the basic map for the purposes of the Turnaround. Typically these will include, but not be limited to:

- any non-load-bearing surfaces, where the siting of cranes etc. is prohibited;
- any areas or roads where access is prohibited;
- approved vehicle routes, with the direction of traffic flow (especially one way systems);
- areas designated for Turnaround stores and quarantine compounds;
- areas for hazardous substances – catalyst, chemicals, paint, oil, fuel;
- foul laydown areas for contaminated items;
- water washing and chemical cleaning bays (water supply, drainage and access);
- clean laydown areas for new and overhauled items;

Figure 8.3 Example of a simple plot plan (Turnaround overlay)

- laydown areas for large fabrications;
- marshalling areas for scaffolding and insulation contractors;
- parking areas for welding rectifiers, air compressors, electrical generators and fuel tenders;
- areas for accommodation, mess rooms, changing rooms and toilet facilities;
- temporary cabling and piping routes for utilities;
- parking areas for cranes and heavy equipment;
- parking areas for site vehicles;
- parking areas for private cars and buses;
- location of additional temporary safety equipment;
- sites for the Turnaround control, induction and safety cabins.

A special additional category is required if any ground is to be excavated. In this case drawings of the site underground services must be obtained and verified – and the relevant information overlaid on the site map.

In addition to the above, the logistics co-ordinator must mark in any other areas or special features local to the particular plant being worked on. Once this is done the plot plan is complete.

Issuing the plot plan

Throughout the creation of the plot plan the logistics co-ordinator consults with the Turnaround manager regarding its feasibility. When it is completed, he or she submits it formally – for discussion, approval and action – to the

Turnaround manager who will then discuss it with the plant manager and his staff, with the safety officers for the site and for the Turnaround, and with any other interested party. Any resulting amendments are then carried out and the final plan published, copies being issued by the logistics co-ordinator to the following personnel (at least):

- Turnaround manager and area engineers;
- plant manager and area superintendents;
- Turnaround safety officer;
- Turnaround planning co-ordinator;
- site security;
- contractor managers;
- any other nominated key persons.

It is the responsibility of the key personnel to study the plan, become familiar with its features, comment upon any circumstance or feature which may affect its usefulness and then brief their staff to ensure that everyone has a good working knowledge of it. The plans should be displayed prominently at key locations in the Turnaround control offices. Briefing and display are very important, because once the event starts everyone on site must conform to the requirements, restrictions and prohibitions of the plot plan. Failure to do so may cause accidents, delay, over-run or overspend. Segments of the plan may be photocopied to aid other specific planning or execution activities but the copies must not be amended.

Logistics is a large and important element in the overall Turnaround plan. As with all the other elements it requires accurate planning and validation.

Marshalling bulkwork

Bulkwork refers to the many small plant items that need to be repaired or replaced during the shutdown. The complexity here lies in the numbers involved and the different ways the individual items are treated.

Figure 8.4 shows a block diagram plot plan for the movement of valves during the Turnaround. In this particular example the valves are removed from the processing plant and due to the fact that they are contaminated with product they need to be cleaned before they can either be worked on or disposed of.

Each valve may be handled in one of three ways (identified in the figure as A, B and C).

Valve A is a rotational change-out where the existing valve is replaced by an overhauled or new spare and is sent to workshops after the Turn-around to be overhauled so that it can be used as a spare on the next Turnaround.

Figure 8.4 Example of a bulkwork marshalling plan layout

Valve B will be sent to the workshop to be overhauled during the Turn-around and reinstalled in the plant.

As an example, its path is: remove from plant > transport to cleaning bay > decontaminate > transport to clean item bay > transport to work-shops > overhaul/repair and test > transport to overhauled item bay > transport to plant and reinstall.

Valve C is a valve that has reached the end of its useful life and will be scrapped and replaced by a new valve.

On a major process plant Turnaround where there could be over 500 valves of various types (control, isolation, non-return, safety relief etc.) some will be changed out, some overhauled and some repaired. In addition the valves may be sent to a number of different workshops, some of which may be a substan-tial distance from the plant.

To complicate things further, each valve will have a particular date by which it must be returned to the plant in order that it can be installed within the duration of the event.

Then, to all of this, add in pumps, electric motors, flow meters, bursting discs, orifice plates and all of the other items of bulkwork and we can begin to appreciate how complex organizing bulkwork can be.

Bulkwork needs a rational systemized plan to ensure that all of the items involved are under control at every stage of their journey.

9

Cost control

Introduction

Creating a budget for a project is relatively straightforward if the exact work scope is known (and will not change over the life of the project) and the unit costs of resources, goods and services are also known. Unfortunately, this is seldom the case with Turnarounds. The budget will be hedged around with uncertainties (some of which have been detailed in previous chapters), the chief of which being uncertainty about the actual work scope (see Chapter 3) and allowance for contingencies. Cost estimating should therefore be performed (for major events, at least) by an experienced cost co-ordinator.

Why estimate a cost profile?

The purpose of putting together a cost estimate or a budget is to ensure that the estimated costs of the event are known as far in advance as possible. This will allow those responsible (mainly the steering group) as much time as possible to optimize the constraints on the budget – by pricing each major element of the Turnaround and holding costs to a minimum. Additionally, the costs are organized to allow expenditure to be monitored and controlled (see Figure 9.1). This demands accurate costing and this in turn depends upon exercising a number of disciplines, namely:

Achieving work definition of the highest possible quality – given the circumstances surrounding the event. Realistically, not all work can be definitively specified. There are unknowns because the items to be worked upon are mainly concealed from view when the plant is running (which is usually the time when planning is carried out). The planner is sometimes reduced to guesswork based on previous experience (which is not always sufficient). Where uncertainties exist or assumptions have been made, they should be specified.

Closing the work list on a specified date – usually four to six months before the start, for a major event. The line must be drawn somewhere to allow the estimate to be made. If the influx of work requests is never-ending a budget cannot be fixed.

Separate costing of all work requested after the work list closure date. Realistically, work requests *will* continue to arrive after the closure date, due to the fallibility of people and equipment. A separate 'Late work order system' should handle this. The work is planned out only after the work on the approved list has been planned (unless there are special reasons why it should be dealt with earlier and these can only be generated by agreement between the Turnaround and plant managers), and it should be costed separately on a late order budget sheet.

Allocation of a unique cost code to the Turnaround. If this is not done it will be impossible to track down the event's costs.

Turnaround manager approval of every item of expenditure – any number of people may otherwise allocate cost to the Turnaround and control over expenditure would be lost.

The inclusions

Having imposed these disciplines, it is then necessary to define what elements shall be included in the cost estimate. These should include, but not be limited to the following.

Turnaround planning and management – this may be charged in a number of different ways, as a fixed and firm lump sum, as a management fee plus reimbursable hours for staff such as planners, or as totally reimbursable for the complete Turnaround team.

Local labour – covering the plant personnel who will be engaged on the event. Some companies do not include these costs in their Turnaround budgeting, but ignoring them leads to an over-optimistic estimate of the total cost.

Contractors – the costs should include those of all the contractors employed. On a major Turnaround there could be thirty or more, from the large multi-skill companies who will carry out substantial packages of work, through the medium sized ones who provide services such as scaffolding and water washing, down to the very small one-task specialists. Care must be taken that *all* contractor costs are recognized at an early date, especially the hidden elements. (In one actual case a contractor presented the client with a bill for 'disbursement' costs of $135,000 which, although not stipulated clearly during negotiations, were held by the contractor to be implied in the contract. The contractor threatened litigation and the client settled out of court.)

Stock and miscellaneous material – stock costs being those of recognized spares, miscellaneous costs those of all materials bought to repair defects etc.

Equipment purchase and hire – equipment can only be properly identified if work methods are specified in detail. Communications equipment – phones, faxes, photo-copiers, computers, printers etc. – is often overlooked.

Accommodation – for a major Turnaround on a plant with limited permanent accommodation, the cost of offices, accommodation cabins, mess rooms, changing rooms and toilets (together with their cabled and plumbed-in services and drainage) can be substantial (even toilet rolls cost money!), especially if the cabins have to be sealed units that may act as refuges during a toxic release. The cost of the fitments (tables, chairs etc.) must also be taken into consideration.

Utilities – a substantial amount of electricity, water, various gases, oil, petrol and diesel will be consumed. (It all has to be paid for.)

Contingencies – as in planning, so in budgeting. Contingency costs are a grey area in which educated guesses based on previous experience (if there is any) are the most accurate estimates available.

The inclusion of the above elements and (with the exception of contingency costs) the accurate calculation of the cost of each will generate the most accurate budget possible.

The exclusions

There are other costs which are normally excluded from the budget. Whether they are, in any particular case, will depend upon company culture and the prevailing circumstances. They are listed below.

Any project work which will be handled independently – although it may be carried out at the same time as the Turnaround and needs to be integrated into the schedule, the cost is not calculated as part of the Turnaround estimate.

Work requested after the closure of the work list – as explained previously, the work list is frozen before an accurate cost estimate can be made. Any work requested after this, if accepted, forms part of the late work order system, and is costed separately.

Emergent work – whether it is generated as extra work, additional work, or by a change of intent, this should be covered by the contingency element of the budget.

Creating a cost estimate

The cost estimate is built up in three basic stages – the ball park estimate, the proposed estimate, and the final estimate. These are developed over a period of time, the accuracy of the estimate being progressively refined as more information is made available.

Stage 1, the ball park estimate

The first stage is to create a ball park estimate using the initial work list. This is typically done six to nine months before the event and has an accuracy of no greater than plus or minus 20 per cent. It is an early warning system that is used to give the policy team an indication of the likely magnitude of the final cost estimate. For example, if the budget allotted to the Turnaround is $2,200,000 and the initial cost estimate is $2,800,000 then it is clear that even the lower band (−20 per cent) of the estimate, at $2,240,000, exceeds the budget – they are not in the same ball park. The higher band at $3,360,000 is radically more expensive. On the other hand, the initial estimate may indicate that the predicted expenditure is well within the allotted budget. Either way, the steering group knows at the earliest possible time what the final outcome is likely to be and, if this is unsatisfactory, has the time to take remedial action.

The initial budget is built up by using one or more of the following techniques:

Analysis of existing data – such as budget and cost reports from previous Turnarounds on the plant in question or on similar plants. The questions to be asked here are:

- Was it a similar event with a similar work scope?
- Can the budget be re-used as it is, or with minor modifications?
- If not in total, can any elements of the budget be re-used?

Analysing the major tasks on the work list – and determining, from plant records, whether similar tasks have been done in the past. If so, how much did they cost?

Aggregating bulkwork – and applying existing labour rates to them. For example, if there are three hundred valves of various sizes and degrees of complexity, calculate an average time per valve in hours and multiply it by an average hourly rate.

Using any existing prices – some tasks may have already been costed.

Using available norms – some companies have calculated norms – based on an average of many similar tasks done in the past – for different types of work.

Using weekly Turnaround control team costs – these will typically have been calculated and agreed during the initiation phase of the Turnaround.

Applying hourly rates for specialist skill groups – these will readily be provided by vendors.

The actual strategy adopted for putting the initial cost estimate together will depend on how much hard information is available. What if there is none, apart from the initial work list? Then, we use a technique which is based on a reasonable ability to estimate averages.

The quick and dirty estimate

In situations where there is little or no historical data, the following technique can be used. It is based on four principal (and not unrealistic) assumptions, namely that:

1. the initial work list is known,
2. basic man-hours can be calculated,
3. current labour rates are known or can be accurately estimated,
4. on Turnarounds, manpower costs typically represent around 30 per cent of total costs. (If better local data, regarding this ratio, exists then it should be used.)

The steps in the technique are as follows:

1. *Select the most experienced team* – from those who have experience on either Turnarounds or maintenance (this would include the Turnaround planners).
2. *Estimate the man-hours for each of the major tasks* – purely from experience, or from past practice, or by simple calculation (analysing the task into steps and estimating, per step, the number of men times the number of hours required).
3. *Estimate the man-hours for each of the small tasks* – a simpler version of Step 2, usually made easier because small tasks can usually be grouped so that an estimate for one of the tasks in any group can serve for each one of the tasks in that group.
4. *Estimate the man hours for bulkwork* – an even simpler task. A large number of similar small tasks (on valves, bursting discs etc.) can be aggregated and the average time per task estimated. In any particular case, the actual number of tasks is then multiplied by that average time.
5. *Add together all of the man-hours* – from the above three categories to give basic total man-hours.
6. *Multiply the basic man-hour total by a productivity factor* – to get the estimated total man-hours. This factor takes account of non-productive hours and is normally between 1.4 and 2.0, depending upon how effective you judge the local labour and supervision to be.
7. *Estimate (or, if possible, calculate) an average hourly cost for labour* – this will be based on the ratios of specialist, mechanical, instrument, electrical, civil, semiskilled and unskilled labour used.
8. *Multiply the estimated total man-hours by the average hourly cost* – to give the basic manpower cost.
9. *Multiply the basic manpower cost by a factor of 3.3 (or whatever factor is most suitable to local conditions)* – to give the basic overall cost of the Turnaround. The 3.3 figure is based on the assumption that labour costs will be one third of total costs – this may be different in different industries.
10. *Consider the worst case scenario* – estimate the overall cost if all the circumstances and conditions assumed turned out, in practice, to be as adverse as they could be.
11. *Consider the best case scenario* – estimate the overall cost if everything went perfectly.
12. Multiply the overall cost derived at Step 9 by four and add to it the figures estimated at Steps 10 and 11, and then divide the total by six. The new total arrived at is the ball park cost estimate for the Turnaround. Obviously this is a very rough figure, but (i) there may be no other way of arriving at a ball park estimate, and (ii) the accuracy sought is only around plus or minus 20 per cent.

This figure, and its estimated limits of accuracy, should be presented to the steering group as early as possible.

The proposed cost estimate

As the preparation phase proceeds and more hard information is obtained on costs, the estimate is refined – using actual and calculated costs – to the point (typically two months before the event) where it can be presented to the steering group as a proposal with an accuracy of about plus or minus 5 per cent. In preparing an accurate cost estimate it must be ensured that:

- every known relevant item of cost is included;
- unit prices are up-to-date and accurate;
- either contingency allowances, for an estimated amount of emergent work, are included or, if not, the exclusions are clearly stated;
- all known or estimated non-productive time is factored in;
- all assumptions are stated.

One of the best investments the Turnaround manager can make is to engage (as suggested earlier) the services of a cost co-ordinator or quantity surveyor to put together the cost estimate. This individual can:

- produce an accurate cost report;
- examine contracts for 'hidden costs';
- challenge work practices to reduce costs;
- present the cost estimate in detail to the steering group;
- create measuring tools for monitoring expenditure;
- measure performance and expenditure before and during the event;
- produce regular live reports and forecasts to allow the Turnaround manager to track costs on a daily basis.

The approved Turnaround budget

When the proposed cost estimate, based on the approved Turnaround work scope, is finalized it is presented to the steering group for analysis, discussion, decision and action (see also Chapter 6, Optimizing the Turnaround plan). If the estimate is less than the allotted budget figure, the saving is quantified and formally recorded. If the cost is greater, the excess is quantified and the following options are explored in order to bring the costs back within the budget figure:

- eliminate work from the work list or reduce the complexity of certain tasks;
- defer some tasks to a later date or another planned outage;
- challenge the unit costs of labour etc.;
- reconfigure Turnaround elements.

The alternative to the above is that the steering group increases the budget. When the estimate is finalized and approved by the steering group, it is adopted as the approved budget for the Turnaround.

During the event, the key members of the Turnaround team are given specific responsibilities for controlling costs. Figure 9.1 shows a check list of actions necessary to ensure that the reponsibilities are met.

Check list of actions required to control costs

Turnaround manager

- brief key personnel on the budget and cost objectives
- set area engineers a cost saving initiative
- agree a cost review timetable with the QS
- vet all variations to contract and approve if justified
- vet variations to labour levels and approve if justified
- vet overtime requests and approve if justified
- vet changes of intent and approve if justified
- vet contractor claims and approve if justified
- review extra and additional costs with area engineers and ensure they are separately recorded
- audit area engineers' cost performance
- review cost report and forecast with QS and engineers and take action to correct cost over-run trends
- report cost issues to the policy team

Quantity surveyor/cost co-ordinator

- produce a cost profile and publish approved budget
- produce regular cost report and forecast, discuss with manager and give early warning of negative trends
- produce financial instructions for contractors
- check contractors daywork claims against contract
- check work measures on scheduled rate contracts
- assist manager to settle contractor claims
- advise area engineers on changes to reduce costs
- check hire dates of all major hired equipment
- collect and collate data from all cost centres
- audit site work to ensure best practice and no waste
- report any cost generating issues to the manager
- write a final cost report with recommendations for future cost saving initiatives

Area engineers

- study the area plan and cost profile to gain a deep understanding of the distribution of costs
- brief the area team on the budget and cost objectives
- control manpower levels and reduce if possible
- vet hire extensions and approve if justified (on major items, check with the manager)
- investigate overtime claims and refer to manager
- investigate daywork claims and authorize same day
- investigate contract variations and record them
- investigate extra work costs and refer to manager
- investigate additional work requests and their cost implications and refer to manager
- review cost report and forecasts with manager and QS
- proactively search out cost saving opportunities

Co-ordinators and supervisors

- do not carry out any extra or additional work on verbal request – ensure it is written down
- do not verbally request extra or additional work from contractors – write it down and report it to the area engineer on a daily basis
- log the use of additional material and hire equipment
- log 'off-hire' dates for equipment and ensure they are met or bettered (under normal circumstances)
- request overtime only when it is unavoidable
- keep tight control of material issued in their area to eliminate waste – which attracts costs
- keep tight control of equipment in their area to avoid damage – which attracts costs
- provide any information requested by the QS

Engineers running contractors

- issue ITBs early to give contractors enough time to develop a realistic price
- ensure ITBs contain all data necessary to allow contractors to identify all work required
- ensure the site is ready for the contractor
- ensure the contractor arrives and leaves to plan
- strictly control all variations to plan and cost them
- minimize contractor waiting time
- report conflict with other work to the turnaround manager as soon as it occurs
- report any contract overrun to the turnaround manager at the earliest possible date
- ensure contracts are properly closed out and final price known at earliest possible date

Turnaround workshop manager

- ensure he has an approved work list
- provide unit prices for standard tasks
- ensure any extra/additional work is logged and approved by the turnaround manager
- if overtime is required, it must be approved by the turnaround manager
- if any task is taking longer than planned, inform the turnaround manager
- provide a daily record of costs for the turnaround manager

Figure 9.1 Cost control

Case study

A chemical company ran three Turnarounds on adjacent plants at the same time. It appointed an event manager for each plant and a Turnaround manager to take overall control of the total project. The latter was given special responsibility for creating an integrated cost estimate, one which would include and combine the costs for all three plants. The company operated a computerized maintenance management system (CMMS) and there was only a limited number of 'experts' in the company who could use it with any degree of competence. There were no fewer than forty two people who had the authority to input work and material requests into the system during the normal operation of the plants. Most of the material for the Turnaround had been pre-ordered under a number of different cost codes and much of it was mixed in batches with materials for routine maintenance.

The event managers did not concern themselves with trying to control costs because they saw this as the Turnaround manager's job. The latter could not stabilize the cost estimate because the costs were lodged in so many different places and obscured by being mixed in with other costs. The company seconded a number of the CMMS experts to help the Turnaround manager to interrogate the system.

The cost estimate was never stabilized. It would rise sharply as a result of one of the CMMS experts suddenly discovering large costs which had been hidden within the system. Then it would fall when other costs were found to be incorrectly allocated to the Turnaround. This situation continued throughout the preparation phase so that the final cost estimate for the events was fictitious. The situation also continued throughout the execution phase and was further complicated by the additional cost of a large amount of emergent work and the decision to take some planned work out of the events in order to try to 'balance the books'. Six months after the Turnaround was completed the final bill had not been calculated because claims for payments were still coming in and there were several disputes between the plants as to who should bear which costs.

What were the basic errors committed in this case?

The senior management did not have a policy for cost control. They assumed that everyone would know what to do and would co-operate in controlling costs. The Turnaround manager did not operate a single cost code system for the event and as there was no penalty for ordering materials without approval, many people did just that. The event managers acted irresponsibly by not controlling their own costs. Ultimate responsibility lay with the policy team who should have addressed the cost issue at their first meeting.

10

The safety plan

Introduction

A Turnaround is a hazardous event. It introduces a large number of people into a confined area, to work under pressures of time with hazardous equipment. In recognition of the greater risk of loss, the targets set for safety on a Turnaround must be uncompromising – zero accidents, incidents, fires etc. In order to meet these requirements the system of working must be equally uncompromising in its approach to safety.

It is the responsibility of the safety officer, leading a team made up of key personnel, to produce a safety plan which will ensure that all relevant matters are addressed. It would be beyond the scope of this book to enter into a full discussion of the nature of such a plan (the documentation for which can run to over a hundred pages for a major event). The aim here is to define the critical items that come within the purview of the Turnaround manager. To begin with, however, it would be useful to define the term 'hazard'.

What is a hazard?

A hazard is any condition, act or event which exposes people, property or the environment to some form of loss, i.e. to:

health	– exposure to any substance, noise or other element which causes temporary or permanent illness or impairment of any bodily function;
life and limb	– any accidental or planned occurrence which causes death or injury to the human body;
property	– any accidental or planned occurrence which causes destruction of, or damage to, plant, structures, buildings, equipment or personal property;
environment	– any accidental or planned emission which pollutes the atmosphere, land or waterway in the vicinity of the plant and beyond.

Factors which contribute to exposure to hazards and consequent loss can include:

lack of proper planning and preparation	– not thinking the job through;
lack of awareness	– ignorance of the hazards;

lack of care and attention – knowing the hazards but ignoring them;
incorrect motivation – putting productivity before safety.

Management of safety is not an academic exercise. It is about real people. The effectiveness of the safe system of work will, in the extreme case, determine whether the people who entrust their safety to that system live or die.

In order to minimize the risk of loss, the Turnaround manager has a number of safety strategies that he can call upon. Also, there are basic principles adherence to which will serve to protect people, property and the environment. The most fundamental of these is the safety chain.

The safety chain

There is a chain of responsibility which runs vertically through the Turnaround organization and its sole purpose is to ensure that those who form its links understand what their responsibilities are and the resulting actions that have to be carried out. The links in the chain are:

Manager > Engineer > Supervisor > Worker.

Manager

The Turnaround manager must provide a safe working routine and ensure that everyone who is employed on the event is properly briefed on that routine before they work on site.

Engineer

The engineers who are responsible for managing areas of the Turnaround must analyse those areas, and the tasks that are to be performed in them, in order to expose, and either eliminate or guard against, any hazards associated with the tasks. The engineers must also ensure that the teams working on the area are briefed daily on safety.

Supervisor

The supervisors must ensure that the workplace is safe at all times, tools and equipment being used are fit for the purpose, and those who carry out the tasks are competent to do so. The supervisors must control the permits in their areas and carry out daily safety briefing.

Worker

The workers must follow all safety instructions and not commit any act, or work in any set of conditions, that would endanger their own or anyone else's life. They must report all unsafe acts and conditions to their supervisor or the safety team so that they may be eliminated.

There are many more safety principles and systems relevant to a Turnaround, but the safety chain is the backbone of safety. Note that the safety officer is not included. This is because safety is the responsibility of line personnel – *not* the safety department. The safety chain forms an important part of the overall safety communications network.

The safety communications network
(see Figure 10.1)

Because of the complexity of the event, it is essential that clear lines of communication on safety requirements are established. Figure 10.1 gives an indication of how many people are involved and just how complex the communications network – the overriding purpose of which is to ensure the safety of the event's workforce – can be. The central spine of the diagram shows the safety hierarchy responsible for setting the safety policy and ensuring that everyone adheres to it. The safety chain is defined, as are the individuals, teams and organizations who can influence the Turnaround. The other functions on the diagram, involving the safety team, will be dealt with later.

Safe system of work (see Figure 10.2)

When the Turnaround manager, or any person nominated by him, sends workers onto a plant to carry out a task, he must ensure *beforehand* that the hazards associated with the task have been taken into consideration. He must then ensure that steps are taken to eliminate the hazards or to protect the workers against them. There are two categories of hazard on a Turnaround, viz:

- a pre-existing hazard, so called because it is inherent in the plant, e.g. hot pipes or toxic substances;
- an induced hazard, so called because it is introduced by the very act of performing the task – e.g. by the use of burning torches.

The safe working routine assesses the following elements:

- the permit to work;
- the workplace environment;
- the worker;
- the task specification;
- materials and substances;
- tools and equipment.

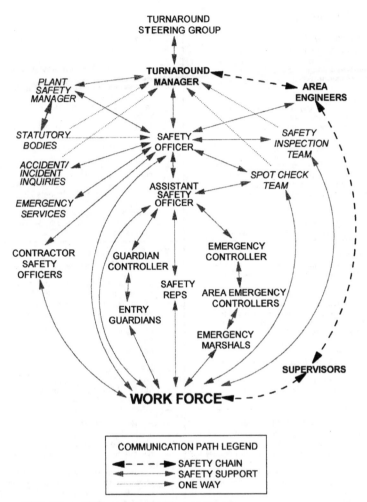

Figure 10.1 The safety communications network

Figure 10.2 offers a number of checklists – which are indicative, not exhaustive – for applying the safe working routine. On one of the above elements, the task specification, a specific hazard assessment must be carried out.

The hot spot inspection

The hot spot inspection deals with pre-existing hazards and is concerned with making the site safe for the Turnaround workers. Well before the event starts, the safety officer and the Turnaround manager should carry out an inspection

Check list of required activities

Permit to work

To ensure safe working a permit must:

- be issued for all tasks and cross referenced to WO
- be specific and relevant to the task being done
- be issued for a specific period of time
- specify, in detail, the work to be done
- specify known hazards and preventive measures
- specify checks to be done and who is to do them
- state if isolation procedure is adequate
- be issued and received by competent people
- be properly handed over at shift change
- be properly handed back on completion of task

Supervisors must spot-check to ensure permit conditions are being adhered to by workers

Safe workplace environment

A supervisor must check the workplace to ensure:

- access and egress safe and adequate for the task
- area is free of dirt, debris and spillage
- area is not contaminated by any substance
- atmosphere is free of dust or noxious substance
- temperature and noise within approved safe levels
- work above/below/beside the work area is safe
- enclosed spaces are adequately ventilated, illuminated and there is an escape route
- where an unacceptable environmental condition cannot be eliminated it is guarded against
- where an unacceptable environmental condition cannot be guarded against – the job is stopped

The worker

The supervisor must ensure that the people who are going to carry out the work:

- are trained and experienced at doing it
- are briefed on safety requirements
- are briefed on the requirements of the task
- understand the permit to work
- understand the task that is to be done
- have the correct materials, tools and equipment
- have been asked for any safety concerns
- feel confident about doing the work
- are wearing the correct protective clothing
- are supplied with safety equipment specified in the permit to work

The task specification

The task specification must:

- specify a safe method for doing the job
- not assume local knowledge
- be easily understood by those who use it
- specify correct material, equipment and services
- identify known hazards associated with the task
- specify safety precautions for known hazards
- include reference to good housekeeping and care for the environment
- identify the use of hazardous substances
- specify decontamination of foul items

Materials and substances

- always ensure that any blanking plate is the correct material for the process fluid involved
- ensure that only specified materials are used
- do not specify or use hazardous substances unless it is unavoidable
- if hazardous substances are used, personnel involved must be properly protected
- before using any new substance, analyse it for any potential hazards
- keep a current register identifying substances hazardous to health (COSHH regulations)

Tools and equipment

- ensure all tools and equipment are undamaged, serviceable and adequate for the task
- discard any tool that is damaged during the task
- ensure air tools have the correct hoses and fittings and that they are periodically inspected
- the use of air equipment to pressure test is forbidden unless written permission is obtained from a competent senior manager
- be aware of the hazards involved in large heavy crane lifts and pre-plan the lift
- ensure that lifting equipment (including beams) has current test certificates

Figure 10.2 Safe working routine

to expose any existing hazards (some of which will not be considered hazards during normal operation but will become so because of the nature of the event). The inspection seeks to expose any condition which could pose a hazard during the event. Hot spots could be:

- non-load-bearing surfaces;
- open ditches and drains (especially beside roads);
- plant which will remain live during the event;

- corroded or damaged structures (especially platforms);
- oily or greasy walking surfaces;
- tripping hazards;
- product drips or emissions (especially acid or toxic substances);
- airborne particles or substances (potential eye injuries);
- restricted access or egress;
- protrusions on walkways (especially at a height);
- ineffective illumination of enclosed structures;
- damaged insulation (especially if it is asbestos based);
- narrow roads (could make transportation or siting of cranes hazardous);
- any other observed unsafe condition.

NB The above list is not intended to be exhaustive.

As a result of the hot spot inspection a list is drawn up of tasks (erection of barriers, cleaning of surfaces etc.) needed to eliminate as many of the inherent hazards as possible. Any hazard which cannot be eliminated must be marked up (in red) on the plot plan as a hot spot. The list of hot spots should be transmitted to the plant team and, in particular, to the permit to work issuers who should be requested to write in special instructions to guard against the hot spot hazards.

Job safety analysis (JSA) (see Figures 10.3 and 10.4)

This subject has already been discussed in the chapter on planning, but it deserves repeating here. During the Turnaround, the worker will interact with all other elements defined in the safe working routine through the medium of 'the task'. It is therefore incumbent upon the responsible person (in this case the preparations co-ordinator) to assess the task in a formal manner. The assessment is, in reality, an investigation of an accident before it occurs, a 'pre-mortem' if you like. It is used to predict potential hazards on a job, to define the types of loss that could result and to specify precautions to be taken to eliminate or guard against those hazards.

JSA is a team activity because one person, working on his own, will consider the job only from his own perspective. The types of task assessed could be major ones, multi-level, complex, unfamiliar, or those with a history of accidents. It is also sensible to carry out a generic assessment of bulkwork. The assessment process has four steps:

1. Specify the main steps of the task.
2. Identify the hazards associated with each step.
3. Define the potential loss associated with each hazard.
4. Detail the precautions necessary to protect against the hazard.

Figure 10.3 shows each of the four steps in detail. Figure 10.4 is an example of a completed task hazard analysis (THA) proforma.

Detailed steps in the task hazard assessment

1. Define the main steps

- select a team with a mixture of engineering, safety, and local plant knowledge and experience
- list the tasks to be assessed from the following:
  - major tasks
  - multi level tasks
  - complex or unfamiliar tasks
  - tasks with a history of accidents
- brief the team on each task
- select the first task to be assessed
- list the basic steps of the task
- check and recheck the steps with the team
- write the steps down on the THA proforma

Note: When developing the basic steps use enough detail to clarify the step but not enough to obscure it. This is best done by following the natural steps of the job

2. Define associated hazards

Does the basic step involve:

- multi level working?
- multi task working?
- working at a height?
- working in an enclosed space?
- loosening, unbolting, moving or lifting?
- welding, burning, grinding, blast cleaning?
- work with damaged or corroded items?
- demolition, generating debris or dust?
- removal of scaffold, walkways or supports?
- noisy, cold, hot, wet, dirty or tiring conditions?
- proximity to live plant?
- complex or unfamiliar activities or equipment?
- any natural environmental hazards?
- interface with public roads, traffic of facilities?
- interface with other plants or companies?

Define the associated hazards for each basic step and record them on the THA proforma

3. Define the potential loss

For each hazard, define the potential loss

Can any person:
- slip, trip, overbalance, fall over/through anything?
- get trapped in, between or under anything?
- be struck or contacted by anything?
- have anything fall on them?
- strike against/come into contact with anything?
- absorb (through skin) inhale or swallow toxins?
- be exposed to radiation, gas or corrosive fluids?
- be burned, electrocuted, frozen, drenched or covered in dust?
- cause a fire or explosion or other destruction?

Is plant or other property in danger of being burned, damaged or wrecked by explosion?
Will any emission caused pollute air, land, stream, river, sea or ocean?
Record potential loss for each hazard on the THA proforma

4. Specify precautions

Prevent the potential loss, i.e.:
- eliminate the hazard
- eliminate the job step
- use an improved method

If the potential loss cannot be prevented, then guard against it, i.e.:
- erect extra supports or scaffold
- erect protective barriers
- post warning signs
- use special protective clothing/equipment
- employ specialist contractors
- limit exposure time
- reduce frequency of tasks
- give personnel extra briefing

If the hazard cannot be eliminated or loss prevented, do not do the job!!

Record the precautions on the THA proforma

Figure 10.3 Job safety analysis

1. Task step	2. Associated hazard	3. Potential loss	4. Elected precaution

Figure 10.4 Task hazard analysis proforma

The safety team (see Figure 10.5, and also refer to Figure 10.1)

The size and composition of the safety team will reflect the company's commitment to safety. If the size of the team or the blend of skills is not adequate then the risk of loss could be increased. This is a matter of judgement for the steering group but the legal requirements must be met.

The safety team has the responsibility for driving the safety message throughout the Turnaround, monitoring safety performance and giving advice and counsel on all matters relating to safety. For a large hazardous event, the company would need to decide which of the following functions it chooses to employ:

Safety officer: a senior person with safety qualifications or safety experience gained over a long period of time. Appointed some months before the event. Translates the objectives of the steering group into a safety policy and writes the safety plan for the event.

Assistant safety officer: normally a supervisor who has experience in safety matters and has the ability to administer the team. Appointed two to four weeks before the event. Takes care of the day-to-day running of the team during the event and is on hand to advise on safety.

Safety representatives: selected from workers who have a specific interest in safety. A mixture of plant and contractor personnel. Appointed a few days before the event and start working on Day 1.

Contractor safety officers: provided by the main contractors and any other contractor who has enough men on site to warrant such a presence. On a large event may be part of the contractor preparation team. If the alliance between the contractor and client is a close one, may act as Turnaround safety officer on the client's behalf.

Turnaround emergency controller: selected, on the day, from plant based personnel who understand the operation of the plant emergency and toxic procedures. Responsible for taking control during an emergency. Normally has other duties.

Area emergency controller: selected, on the day, from engineers or superintendents (client's or contractor's) to be responsible for the men on their areas during an emergency when they will work under the direction of the Turnaround emergency controller. Have their normal duties to perform.

Emergency marshals: selected, on the day, from trade supervisors with breathing apparatus training. They will assist the emergency controllers to marshal people during an emergency and form the search and rescue teams if anyone is unaccounted for.

Entry guardian controller: selected, a few days before the event, from the safety representatives. Responsible for the personnel who act as guardians for the protection of workers covered by entry permits.

Check list of responsibilities

Safety officer
- manage safety team and report to the turnaround manager via daily turnaround control meetings
- formulate the turnaround safety plan
- organize all turnaround safety inductions
- drive safety campaigns and initiatives
- organize daily site safety inspections
- organize random 'spot-checks'
- publish daily safety reports
- organize a safety suggestion/complaints book and analyse it on a daily basis
- make recommendations for changes to safety policy/procedures to improve safety

Assistant safety officer
- supervise safety team on a day to day basis
- report to safety officer
- monitor safety representatives performance daily
- monitor emergency team's performance daily
- provide safety advice to all personnel
- respond to any unsatisfactory safety condition, accident, incident or emergency
- solve first line safety problems and refer those outside scope to the safety officer
- keep all safety records and statistics
- provide information for daily safety report
- assist safety officer with safety inductions

Safety representatives
- represent the entire work force
- constantly monitor work areas to drive safety
- inspect, scaffold, platforms, accesses, barriers
- control housekeeping and general plant state
- monitor wearing of safety clothing
- monitor traffic movement on site
- monitor work practices
- record all safety infringements
- check entry guardians are in place
- take part in 'spot-checks'
- provide safety advice to all personnel

Contractor safety officers
- ensure contractor employees conform to the safety requirements agreed in the contract
- ensure all contractor employees receive safety induction and are issued with a security pass
- ensure no contractor employee is required to carry out any task he deems unsafe
- record and report any safety infringements
- attend the daily turnaround safety meeting
- if required, attend the daily control meeting

Turnaround emergency controller
- manage the turnaround emergency system
- manage the area controllers and marshals
- establish links with emergency services
- monitor day to day running of the emergency system run by area controllers and marshals
- manage all fire, toxic and incident alerts
- report on a daily basis to the safety officer
- carry out random checks on toxic refuges
- carry out random spot checks on fire equipment and assembly points
- must be contactable at all times while on site
- must deputize an area controller to cover at all other times – including night shift

Area emergency controllers
- manage the emergency system in a specific area
- co-ordinate and monitor the daily performance of the emergency marshals
- collate the daily 'refuge sheets' and report results to turnaround emergency controller
- ensure that all emergency marshals have the required safety and emergency equipment
- organize emergency marshals during a site emergency (fire, toxic or major incident)
- report to turnaround emergency controller on a daily basis
- must be contactable during the shift
- must have a deputy to cover for absences

Emergency marshal
- responsible for people accommodated in cabins or rooms designated as 'emergency refuges'
- ensure all personnel sign in and out on the emergency register on a daily basis
- call the roll in the cabin during a toxic or other emergency requiring men to seek refuge
- call the roll at assembly point during fires
- report roll calls to area emergency controller who will contact with missing person list
- search for missing persons during emergency
- lead first line fire fighting teams
- must be contactable during the shift
- must have a deputy to cover for all absences

Entry guardian controller
- manage the people selected to act as guardians for work teams entering enclosed spaces
- ensure the appropriate persons are properly equipped and are in contact at all times – contact them regularly
- respond to any distress call from the guardian and inform the assistant safety officer
- visit the guardians regularly to ensure they are performing their duties and not being harassed
- if an accident occurs, inform the safety officer, turnaround manager and emergency services
- at times of accident do not leave post to assist – be the vital communication link

Figure 10.5 The safety team

As well as the specific responsibilities of team members, as explained in Figure 10.5, there are a number of other matters concerning the team, namely:

Safety cabin: the team is housed in a safety cabin. On many Turnarounds this is painted in distinctive colours (i.e. green and white stripes).

Safety log book: in the cabin there is a log book in which any person on the Turnaround is free to report unsafe acts and conditions or make suggestions for improving safety.

Newsletter: it is common practice on Turnarounds for the safety team to publish a daily safety newsletter to communicate all matters relevant to safety to the widest possible audience.

Safety incentive schemes: a rather vexed subject. Some companies believe it to be a good way of focusing the workers' minds on safety, while others contend that it is counter productive (for example, if awards are made for accident free days it may encourage an individual to hide the fact that he has been injured in order not to 'spoil' his colleagues' chances of winning the award). Other companies insist that safety is a statutory requirement, and therefore people should not be rewarded simply for doing what they are supposed to do anyway. The company culture determines whether or not safety awards are made and, if so, what form they will take. If incentive schemes are adopted, it is normally the responsibility of the safety team to administer them.

So far it is the preparation for safety that has been discussed, and the systems and routines put in place to minimize loss. The other, equally important, aspect of the safe system of work is monitoring safety performance.

Safety inspections (see Figure 10.6)

On Turnarounds, the monitoring of safety performance is achieved by the practice of two types of formal safety inspection: daily, of the general area, and spot checks on specific jobs.

The daily safety inspection (see Figure 10.6)

The daily inspection is carried out by a team drawn, if possible, from a list of directors and managers, who have been invited from all parts of the company and have agreed to take part in a rota for carrying out such a duty – on dates specified on a programme drawn up by the safety officer. It is a given in any safety programme that the commitment of directors and senior management is crucial to its success. By participating in this way, the senior staff of the company are seen to demonstrate their commitment.

The duration of the inspection is normally one to two hours. The managers meet the safety officer beforehand, who briefs them on what is expected of the team. The inspection should have a daily theme (e.g. the wearing of light eye protection) as well as carrying out the general search for unsafe acts and conditions. When it is completed, the team is de-briefed by the safety officer and this generates a report concerning any occurrence of

Check list of activities

The daily safety inspection

The team follow a check list such as the one below

Daily safety theme
- is everyone aware of the theme. If not, why not?
- are supervisors enforcing the safety theme?
- are workmen complying with the safety theme?
- is anything making the theme difficult to enforce?

Unsafe acts
Is there a risk that any worker will endanger himself or others because he is working:

- with improper motivation?
- in a dangerous location/prohibited area?
- in dangerous conditions?
- with dangerous substances?
- with dangerous equipment (or misusing, abusing tools or equipment)?
- with an unapproved job method?
- without proper PPE/protection/precautions?
- without proper supervision, or alone?
- in a confined space without a guardian?
- in any other dangerous way?

Unsafe conditions
- is any permanent or temporary structure unsafe?
- are there holes/gaps on platforms or handrails?
- are there any unsafe or damaged ladders, small platforms, stairways or zip-up scaffold?
- is any live pressurized or high temperature or moving plant inadequately guarded?
- is radiography taking place without proper isolation/barrier/sign – above/below/adjacent?
- are vehicles being driven recklessly or onto areas where they are prohibited?
- are any roads, paths, walkways or stairways blocked by equipment or materials?
- is any dangerous projection unguarded?
- are any teams working inside vessels prone to poisonous, pyrophoric or electrostatic build-up?
- is there any precaution which protects plant but places people in danger?
- is there any other unvsafe condition?

Housekeeping
- is the workplace dirty, oily, greasy or wet?
- is there debris or rubbish lying about?
- are rubbish skips inadequate or hidden?
- are there insufficient plastic rubbish bags?
- are nuts, bolts, washers etc. being left lying on platforms or on the ground?
- are supervisors not enforcing good housekeeping?

The spot check

The purpose of the spot check is to ensure work teams are working safely and in safety

Spot check technique
- list a number of tasks for consideration
- select one task at random from the list
- study the task specification sheet and related documents to familiarize the team with the task
- go to the process cabin and inspect the master copy of the permit to work (photocopy it)
- visit the work site and observe work conditions and team behaviour
- announce a 'spot-check' and stop the job safely
- discuss the permit to work, the task and the working conditions with work team members
- restart the job when it is safe to do so
- record and report all safety infringements to the safety officer and turnaround manager

At the end of the spot check the team should be able to answer the following questions:

- is the area engineer managing his area properly?
- is the co-ordinator providing all that is required to ensure job safety?
- is the supervisor in control of the job?
- has the work team been properly briefed on the task and on safety requirements?
- do individuals know what to do in an emergency (fire/toxic/accident/incident)?
- is the permit to work on at the job site identical to the one in the process office. If not, why not?
- is the permit adequate?
- do the work team understand the permit, the hazards and the precautions?
- are the team complying with precautions?
- is the task specification safe?
- are the team complying with the task specification?
- are the team using the correct tools and equipment and are they using them safely?
- are individuals wearing the correct safety gear?
- are all access, egresses and platforms clear and properly maintained (check scaff-tag)?
- is there any atmospheric condition which could endanger the health of the team?
- are there any substances which could endanger the health of the team?
- is there any condition in the work area which could endanger the safety of the team?
- is there any other safety issue worth recording?

Figure 10.6 Safety inspections

unsafe acts or conditions that the team may have found. The safety officer relays this information to the area engineers who are responsible for eliminating the unsafe acts or conditions. The safety team regularly monitors the areas to ensure this is done.

The spot check

Spot checks (sometimes referred to as 'job freezes') are carried out on a random basis by a team selected by the safety officer. They are usually of one to two hours' duration. The team meets for briefing beforehand and for de-briefing afterwards in the same manner as for the daily inspection. However, the focus is the opposite of the daily inspection, in that it is aimed at one specific task. The check tests whether the people working on the task are aware of the safety requirements and are conforming to them. The right-hand column of Figure 10.6 details the protocol of the spot check.

If faults are found they need to be recorded, rectified as quickly as possible and the rectification checked by the team who carry out the next inspection.

Unfortunately, even with all of the safety systems in place, accidents or incidents which have to be investigated may still occur.

Investigating accidents (see Figure 10.7)

Defining the term 'accident'

An accident is an event which causes death or injury to people, damage to property or pollution of the environment. Accidents can be minor, serious, major or catastrophic.

Why investigate?

Accident investigations are carried out:

- because it is the responsible and caring thing to do;
- because it is required by law;
- to eliminate the recurrence of similar accidents;
- to reduce the anxiety of workers who are emotionally affected;
- to understand the root causes of accidents;
- to provide hard data for feedback to the safety system;
- to provide information to a higher authority.

If an accident does occur, an inquiry is convened by the Turnaround manager to establish the facts about the five links in the accident chain, namely:

1. *loss* – the type and magnitude of loss in terms of death or injury, damage and environmental pollution;
2. *incident* – the specific incident that caused the loss, in terms of contact with a substance, object or energy source;

Inquiry attendees

The inquiry will involve some, or all, of the following:

Members
- Turnaround manager or nominee (Chairman)
- Turnaround safety officer (Secretary)
- Plant safety manager (if applicable)
- Plant manager or nominee
- Process supervisor
- Maintenance supervisor
- Selected safety representatives

Witnesses
- Turnaround area supervisor involved
- Persons involved (if available) and witnesses

Others
- Management or trade union observers
- Legal representatives

Establishing the facts

Reports are read, documents studied and people questioned to establish the facts of the accident and in some cases it may be necessary for the team to visit the site of the accident to establish:

1. *Loss*
- how many people were killed or injured?
- the nature of the injuries of each
- the damage to plant and property
- the nature of pollution of the environment

2. *Incident*
- what object or substance caused the injury in terms of impact, cutting, abrading, crushing etc.?

3. *Immediate causes*
- what made the incident occur in terms of defective equipment, conditions or acts?

4. *Basic causes*
- why did the immediate cause arise, in terms of lack of control, lack of awareness or incorrect motivation?

5. *Loss of management control*
- why was the basic cause allowed to arise, in terms of, inadequate systems, standards, procedures, training or communication?

6. *Recommendations*
- what can be done to regain management control, eliminate basic causes and prevent recurrence in terms of:
- improving systems/standards/procedures?
- briefing/retraining personnel?
- upgrading/improving equipment and plant?
- providing protection or warning signs?

Preparations

To prepare for the inquiry any or all of the following should be obtained:

- copy of the supervisor's first line accident investigation and report
- both copies of the permit to work
- the task specification being used at the time
- any supporting documents such as drawings, weld, inspection or pressure test specifications
- P and ID's, line diagrams and plant drawings
- specifications of all materials and substances
- specifications of tools and equipment used
- any history of similar accidents on the plant
- the names and designations of people involved
- the names of any witnesses
- the facts or an assessment of the loss
- any written report/testimony about the accident

Analysing the inadequacies

Questions to uncover basic causes

Environment
- was the work area safe, in terms of access and egress, atmosphere, configuration?

Procedures
- was the task properly organized, in terms of supervision, instruction, communications task specification, other information?

Equipment
- was equipment adequate, in terms of ergonomics, type, installation, serviceability or protection?

People
- were the people involved competent, in terms of physical and mental aptitude, attitude, training and experience, skill, motivation or stress?

Inquiry report

The report should contain the following information:

1. A summary of the accident/incident
2. Details of injuries or fatalities
3. Details of damage to property
4. Details of environmental pollution
5. The basic causes of the accident/incident
6. Recommendations to prevent recurrence of the accident/incident

Figure 10.7 Investigating accidents

3. *immediate causes* – of the incident in terms of human acts and working conditions;
4. *basic causes* – underlying the immediate causes in terms of human factors and job factors;
5. *loss of management control* – in terms of systems, standards and procedures.

After many years of accident investigation in many industries in many countries, two basic rules concerning accidents have become evident:

1. The specific circumstances surrounding accidents are almost never repeated.
2. The underlying causes of accidents are always the same and are any combination of the following:
   - uneducated, incompetent or uncaring management;
   - inadequate safety systems, standards and procedures;
   - bad planning and preparation;
   - inadequate individual skill;
   - inadequate safety awareness;
   - incorrect motivation.

Safety is the business of everyone connected with the overhaul, but it is the specific responsibility of the Turnaround manager to ensure that there is a safe system of work in place.

Case study

During a Turnaround on an acid plant a large vessel with a dished bottom end, four-metres in diameter, was opened up for inspection. The first person into the vessel – a scaffolder (stager), whose task was to erect a platform for the inspector to stand on – slipped down the dished end and broke his wrist.

The accident enquiry investigated and found the following.

The dished end was made from highly polished (and very expensive) chrome–molybdenum steel. It had a number of stepping bars welded onto the surface to allow a person to climb down from the manhole to the bottom of the dished end but the bars stopped one metre from the bottom of the dished end so the person would have to step out onto the curved surface.

During previous overhauls the scaffolders had damaged the surface by walking on it with rubber boots, possibly contaminated with mud. The then plant manager (five years before) had written a site instruction – which was actually written on the permit-to-work! – that anyone entering the vessel must wear cloth or paper overshoes. He had rightly been concerned about the damage being done to the vessel surface but he had not considered the effect of someone trying to walk on a curved, highly polished surface without anything to hold on to.

In the previous Turnaround (three years before) an identical accident had occurred.

There had been an enquiry which recommended the use of a safety harness coupled to a hand-operated hoist so that the scaffolder could descend under control. The recommendation had never been adopted.

It was also discovered that another nearby plant owned by the same company had similar vessels and had solved the problem by issuing the scaffolders with special boots with non-slip soles that did not scuff the surface. Both the boots and the harness idea were adopted for future Turnarounds.

What are the lessons to be learned here? A failure to consider the safety ramifications of an instruction set this particular accident up. A failure to adopt previous enquiry recommendations caused a repeat of the accident. Useful safety information (regarding the boots) was not circulated around the company and an avoidable accident was allowed to occur. So the lessons are: always consider the safety issues when changing or modifying anything; always adopt enquiry recommendation when it makes the job safer; circulate safety information around the company so that everyone gets to know about it.

11

The quality plan

Introduction

As emphasized in Chapter 1, a Turnaround is a strategic management tool used to safeguard plant reliability. The plant has three basic functions:

- to transform material from one state to another;
- to transport material from one place to another;
- to contain the material during transformation and transportation;
- to control the manufacturing process.

All of the various items of equipment on the plant are configured to facilitate the most effective performance, in line with current knowledge, of the four functions. A Turnaround is performed, as part of an overall maintenance pro-gramme, to protect these functions. However, it is a radical intervention – many parts of the plant are taken apart, worked upon, and then put back together. There is a real risk of introducing unreliability via the very act, which is per-formed to safeguard its reliability. Planning, preparation and execution must therefore be performed in such a manner as to ensure that the integrity of the plant is not adversely affected.

Many companies employ auditors (see Chapter 1 and Part 2 of this book) to examine their performance, highlight any shortcomings and recommend improvements. To assure quality, the requirements of every task must be correctly specified and then performed to that specification. Everything must be done correctly. The way to 'get it right' is to have a coherent, auditable quality trail from initial work request to final acceptance of the completed task. Figure 11.1 is an example of a task preparation checklist (indicative, and not exhaustive). By following it the planner can ensure that all the necessary steps have been carried out, and an auditor can follow the logic.

The type of quality system used for any particular Turnaround, and which will be established by the steering group, will depend greatly upon the general approach to quality within the company.

Steering group issues

Quality – conformance to requirements – is vital to the success of any Turn-around. Unfortunately, if the amount of time, effort and money which has been expended on it over the last decade is anything to go by, its promotion does not

	Yes, No or N/A
01. Has the work request been endorsed by the plant manager or nominee?	
02. Have plant drawings/documents been validated by the plant manager or nominee?	
03. Is issue of above drawings and documents controlled by the planning co-ordinator?	
04. Has the plant manager specified the plant standards and operating procedures?	
05. Has the work scope meeting validated the task?	
06. Have plant personnel tagged the task on site with a unique number (tag number)?	
07. Has turnaround manager specified if task is to be done in house or by a contractor?	
08. Has the planner specified the task on a task sheet or list and tag numbered it?	
09. Has the planner specified materials, with material certificates for critical materials?	
10. Has the planner specified equipment with calibration certificate for measuring instruments traceable to national standards?	
11. Has the planner specified all required services and utilities?	
12. Has the planner cross-referenced supporting documents on the task sheet and defined them on an index sheet to create a task package?	
13. Has task package been approved by maintenance/operations managers or nominees?	
14. Has the planning co-ordinator scheduled the task in the turnaround plan?	
15. Have all materials/equipment/services been procured by logistics team?	
16. Have materials/equipment/services been checked on arrival at site by the appropriate person, i.e. planner, engineers, inspector etc., and:	
16.1 has any item been rejected/quarantined (in writing) by the 'inspector'?	
16.2 has any item been concessed in writing by the plant manager?	
16.3 does critical material have a material certificate?	
16.4 does each measuring instrument have a current calibration certificate?	
16.5 have all rejected items been returned to the supplier and re-ordered?	
16.6 have all items been properly stored and protected?	
16.7 is there a controlled procedure for issuing items from stores?	
17. Has the planner issued a copy of the task package to the supervisor?	
18. Has the planner shown the supervisor the task on site and briefed him on the task?	
19. Has supervisor checked the task tag number against tag number on the task package?	
20. Has the supervisor checked that material/equipment/services needed for the task are available together with any material and calibration certificates?	
21. Has the supervisor ensured that personnel chosen to do the task are competent?	
22. Has the supervisor briefed the personnel on the skill and safety aspects of the task?	

If the above activities conform to requirements then the task may proceed

Figure 11.1 Sample audit trail for task preparation

seem to have been an inherent characteristic of many industrial enterprises. Assurance and control of the quality of any project require a system dedicated to that end. Not having one for a Turnaround can lead to over-run, overspend and inferior work that will compromise plant reliability.

The steering group's approach to quality will depend greatly on the current level of quality awareness within the company. In many cases the group will look to the Turnaround manager to guide them on the question of quality systems. He must investigate the company's existing systems and, depending upon what is found, take one of the courses of action that will now be discussed.

ISO 9002

If the company operates an internationally or nationally accredited quality system such as ISO 9002, the Turnaround manager should adopt it.

Internal quality system

If the company has a well defined set of internal quality procedures, the Turnaround manager should adopt them.

Fragmented procedures

If the company has a number of procedures which cover quality but are not integrated into a system, they should be integrated into one system and, where necessary, augmented by writing additional procedures.

No formal quality system

If no formally written down procedures are followed, the Turnaround manager should ascertain if there is an informal system at work and, if there is, whether or not it is adequate. If it is, it should be formalized by writing simple procedures and, if necessary, augmented by writing additional procedures.

No system

If the company has no formal system and the informal one is inadequate, the Turnaround manager should write out a set of simple quality procedures and check-sheets to cover the Turnaround.

Whichever of the above options is chosen, the Turnaround manager should propose the quality plan for the event to the steering group, who should then discuss the requirements and come to an agreement on what the approach to quality will be. It is the Turnaround manager's responsibility to propose the quality plan, but it is the responsibility of the steering group to approve it.

They must also bear the consequences if there is a failure due to the lack of an adequate quality system.

Basic quality requirements

A lot has been written over the last thirty years about quality. It has been analysed and presented from many different viewpoints. Take these three ideas of quality from three different perspectives

(a) Behaviour – Quality means 'conformance' to requirements

This implies that the people providing the 'inputs' to a system or process need first to know what is required of them via roles and responsibilities, standards and procedures etc. imposed by someone – normally management – and then act in such a way as to conform to them.

(b) Products – Quality means 'fitness for purpose'

This means that the product (or 'output' of our system) should be capable of performing to the specification set for it. For example, if a pump is rated to work continuously for 15,000 hours at a certain flow and it does so then it can be considered to be a quality product. It is no great stretch to translate this into 'conformance to requirements'.

(c) Psychology – Quality is an 'attitude of mind' that drives us always to do our best

This is a more subtle but also much more powerful definition and describes the real basis of a true quality approach. For example, in (a) above, we would conform because we were 'required' to by some external influence (the management?), whereas in this example we would conform to requirements because of an internal drive – to meet the values (requirements?) we set for ourselves. This attitude has been referred to as 'Quality People'.

Many companies have found to their cost that having a quality system and procedures without the buy-in of their employees can generate a futile and costly exercise in bureaucracy.

Regarding the three examples above, at first glance they may seem pretty diverse but they are, in fact, inextricably linked.

For example, if our Turnaround is to be 'fit for purpose' (by achieving its QQTMS objectives) then those people who plan and execute the event need to understand the Turnaround requirements and conform to them, and the people most likely to conform effectively are those who have the internal drive to do so.

The elements of quality

For the purposes of this chapter we will define two elements of quality that are necessary to achieve success on a Turnaround. If we look at the most basic level of a work process, we get

INPUTS > WORK PROCESS > OUTPUTS

And if we remember the GIGO principle (garbage in–garbage out), which states that no matter how good your work process is, if your inputs are substandard then your outputs will be substandard, this also exposes two fundamentals of quality:

1. We must have some way of guaranteeing our inputs are adequate;
2. We must have some way of guaranteeing that our outputs are adequate.

This is the essence of this chapter. We will define two elements, quality assurance and quality control, in a very specific way:

QUALITY ASSURANCE – Actions taken before the event to eliminate faults, and
QUALITY CONTROL – Actions performed after the event to expose and rectify faults.

The systemic connection between quality assurance (QA) and quality control (QC, see Figure 11.2) is that the measures and techniques to be used for QC are determined by QA requirements and the learning from the actual QC perform-ance is fed back to the QA requirements to improve them continuously.

For example, if QA requires a Class 1 weld but sets the inspection require-ment at 'dye-penetrant testing', QC would find it impossible to meet the requirement because a Class 1 weld must be free of porosity and dye-penetrant testing cannot determine this, therefore QC would feed back to QA the infor-mation that the weld inspection needs to be 'radiography', so that in future all Class 1 welds would be radiographed.

Quality consists of two elements – both are required on a Turnaround

Figure 11.2 The elements of quality

Both QA and QC are required on Turnarounds because of the hazardous nature of the production processes that need to be accounted for when looking at the reliability of the plant (high pressure, high temperature etc.).

We will now look at QA and QC in more detail.

Quality assurance

Quality assurance is where the majority of value lies. The effort expended before the event to eliminate faults will be repaid many times over during the event and, as part of the pay-back, will render QC simpler and easier to perform.

Quality assurance on Turnarounds

To apply QA to Turnarounds (see Figure 11.3) we need only fill in the requirements for the event and then check each of them to 'assure' ourselves that they are 'fit for purpose' because their fitness will determine the output of the Turnaround process (GIGO rule applies). Figure 11.3 shows only a sample of the many issues that need to be covered by the QA system to ensure quality.

Typical Turnaround quality assurance issues

Figure 11.4 shows a sample of the many activities required during the planning and preparation of a Turnaround to assure that all QA aspects have been organized to eliminate faults from inputs.

Figure 11.3 Quality assurance for Turnarounds

Figure 11.4 Typical Turnaround quality assurance

Material traceability

Some material (using it in its general sense to include raw materials, proprietary items and consumables), because of its critical nature, may require evidence of source in the form of certification or guarantee. This must be obtained *before* the material is installed in the plant.

Calibration traceability

If tools and instruments that require to be calibrated are used then we must (a) ensure that they are correctly calibrated, (b) ensure that the equipment used to calibrate them has been calibrated and (c) ensure that such calibration is traceable back to a national standard. Otherwise we cannot guarantee that the setting for the tool or instrument is correct.

Case in point – the author once audited a company who were using pressure gauges on their high pressure testing of vessels and pipe work on a Turnaround. They were asked how they guaranteed that the gauges were correct. They stated that they 'tested' their gauges by connecting them to a 'pressure tester' in groups of three. If all three gauges read the same pressure then all three were used. If one read a different pressure from the other two then it was discarded and the other two were used. It does not take much analysis to discover the flaw in this logic!

Remember, if a calibrated tool, let's say a torque wrench, is incorrectly calibrated then every fastening tightened by that wrench will be wrong.

Accuracy of documents

On large ageing plants that have undergone both a number of 'reorganizations' and many 'modifications' without the drawings and documents being amended, and updated their CMMS facilities a number of times and 'lost' data in the process, this issue can assume nightmarish proportions. The drawings and documents are the specifications of our plant. If they are wrong they are the garbage in GIGO. They must be carefully checked to ensure that either they are up to date or there is some responsible person on the plant who will vouch for their accuracy – in writing.

Weld quality

Welders are almost unique among tradesmen in that they have to constantly 'prove' their competence by passing welder qualification tests on a regular basis and, in some cases, before every project. We must ensure that the qualification tests are adequate. It is also important that the weld procedures, which describe how the weld is to be done, are adequate and that the inspection techniques set to check the weld against the standard are the correct type.

Pressure test control

Pressure testing, especially at high pressures, is a potentially hazardous process. The potential for hazard increases greatly if we are using pneumatic testing. This is because of the simple fact that unlike water (used for hydraulic testing), which is not compressible and does not store energy, gas is compressible and does store energy.

This means that if the item being tested fails the pressure test and breaches (causing damage to the item), the outcome for hydraulic testing would be a leak whereas the outcome for pneumatic testing would likely be an explosion. This is why pneumatic pressure testing is so rarely used.

Due to the hazardous nature of pressure testing we must ensure that the equipment we use is serviceable and that the standards and procedures we use are written and approved.

Task control

The actual tasks that are carried out on the Turnaround should also be subject to QA because the information contained in them is the primary input into the Turnaround process. We must ensure that the job specifications are validated before planning starts, the work packages are validated on completion of planning, and the schedule (which will be used to control the progress of the event) is validated before we commence work.

Design verification

Where projects to improve plant performance are included in the Turnaround worklist, they normally constitute a modification to the plant. We must therefore ensure that they are approved under a safe modification system and are technically feasible (at least the installation phase).

Quality plans

The difference between a normal job method and a quality plan is that the latter is written for technically critical jobs and incorporates 'hold points' where inspection *must* be carried out before the job is allowed to proceed to the next activity. At each hold point the level of inspection needs to be specified (client inspection or independent inspector etc.).

This list is not exhaustive and many industries have their own unique QA requirements. Each should be treated with the same care and attention to 'assure' the inputs into the Turnaround process.

Quality control

Includes all activities performed after a task has been completed to expose and rectify faults.

The need for quality control

The overriding issue regarding the need for quality control on Turnarounds is that there is so much of it going on at the same time (pressure testing, radiography, MPI, visual inspections, physical measurement, leak testing etc.). It all has to be organized and executed in the same planned manner as the scheduled work.

Typical Turnaround quality control issues

Figure 11.6 shows some of the QC actions that need to be performed during the event to expose and rectify any faults that have occurred during the execution of the work, i.e.

Figure 11.5 Need for quality control

Actions carried out after the event to expose and eliminate faults

Figure 11.6 Typical Turnaround quality control issues

Pressure testing

To ensure the structural integrity of vessels and pipe work.

Weld inspection

A number of different techniques to ensure that any faults in a weld are within an acceptable predetermined standard.

Alignment checks

On rotating machines, to ensure that, for example, the rotors have been installed correctly within predetermined limits. The traditional method of 'shims and feeler gauges' has, in many cases, been replaced by laser alignment.

Joint inspection

Especially in the process industry where there may be thousands of joints broken and remade during a shutdown. Some of these will be critical, e.g. on pipelines carrying flammable or toxic materials, and require extra layers of inspection to ensure they do not leak.

Configuration check

To ensure that an item has been installed the correct way round. It may seem hardly credible that an item could be installed the wrong way round but there have been occasions, known to the author when pumps have been installed that were pumping against the flow of product!

'Hold point' inspection

This needs to be organized on the day, so that it is carried out without delay to the job, and at the correct level.

End of shift calibration check

This is to ensure that the calibrated tool or instrument used during the shift has the same readings at the end of the shift as it did at the beginning.

Handover quality checks

This refers to any checks that production personnel need to carry out on work done before they accept the handover of equipment from maintenance.

Turnarounds require many QA and QC activities to be carried out to ensure that first, as many as possible faults are eliminated prior to the event and then such faults as do occur are exposed and rectified. Only the twin application of QA and QC is adequate to meet Turnaround requirements

Considering joints

In processing plants, the breaking and making of a joint is, on the face of it, one of the most straightforward of tasks on a Turnaround, but it is also the most frequent one; not uncommonly, many thousands of such jobs will be required. In addition, joints are critical to containment function and badly made joints are one of the fundamental causes of duration over-runs (not to mention the potential hazards caused by uncontrolled emissions). All leaks should be deemed unacceptable but there are some joints which are more critical than others and they, at least, should be subjected to a quality inspection to eliminate, as far as possible, the possibility of leakage on start-up. Figure 11.7 is an example of a joint check sheet. At the very least, the client should specify a list of critical joints which have to be inspected before making.

Case study

Some years ago the author was involved in a Turnaround on a plant which suffered from a large number of leaks on start-up. It had to be brought off line three times, and four days' production was lost due to this. After the event a corrective action team was formed, as part of a quality improvement programme, to try to identify the root cause of the leaks.

The team was made up of mechanical fitters and supervisors and chaired by the author. A brainstorming session was organized to generate as many reasons as possible which might account for a set of two related joints failing

Critical Joint Inspection Routine - *Stripping the joint*

1.   What is the tag number and orientation of the insert (pump or valve etc.) ?

2.   Have any of the joints been filled with resin, if so, how will the resin be removed ?

3.   When the resin is removed, is there any evidence to suggest what ma have caused the leak ?

4.   Are the pipe and insert flanges the 'correct' schedule ?

5.   Is there an excessive cold gap between the pipe and insert flanges ?

6.   Does either pipe 'spring' when unbolted - due to axial misalignment ?

7.   Are the bolt holes in the two pipe flanges 'off pitch' - due to radial misalignment ?

8.   Is there - scoring, gouging or pitting on either of the flanges ?

9.   Are the flange faces correctly machined - plain, stepped, serrated etc. ?

10.   Have either of the flanges been over machined - to below acceptable wall thickness ?

11.   Is the gasket being removed from the joint the correct one ?

12.   Is the gasket damaged, indicating there is something wrong with the joint ?

13.   Are the fasteners used on the joint of the correct type, and are they undamaged ?

14.   Are the fasteners to be reused, if so, how will they be protected until required ?

15.   How are the two 'open' pipe ends to be temporarily sealed until the joint is reassembled ?

The results of the above survey should be reported to the Area Engineer so that action taken to eliminate any faults.

Critical Joint Inspection Routine - *Remaking the joint*

1.   Was the joint inspected when it was stripped - if not, why not ?

2.   Have the necessary refurbishment and machining actions been performed correctly ?

3.   Have either of the flanges sustained any damage since the joint was broken ?

4.   Has the correct insert been replaced in the correct orientation ?

5.   Have the two mating pipe flanges been aligned (axially and radially), without the use of force ?

6.   Are the gaskets correct and free from faults ?

7.   Have the insert and the gaskets been reassembled in position, without the use of force ?

8.   Have the fasteners been passed as fit for purpose and are they properly lubricated ?

9.   Have the fasteners been inserted in the bolt holes, without the use of force ?

10.   Have the fasteners been tightened in the correct sequence ?

11.   If required, have the fasteners been tightened to the specified torque, using calibrated equipment ?

12.   If the joint has been 'flogged', was it performed by an appropriately experienced craftsman ?

13.   Is the joint acceptable - if not, state the reason ?

Any concerns with regard to the quality of any joint should be reported to the Area Engineer so that appropriate action can be taken.

Figure 11.7 Joint check sheet

by leakage. No less than sixty three different possible factors were postulated, among which were:

- radial and axial misalignment of the two associated pipes;
- incorrect cold gap between flanges prior to tightening;

- poor flange face quality;
- flanges thinned due to previous over-machining;
- poor gasket quality;
- poor quality of joint assembly;
- employment of incorrect bolt tightening sequences;
- inadequate competence of the technician doing the job;
- poor quality of inspection or monitoring;
- the integrity of one set of joints being disturbed by the inferior assembly and tightening of another set of joints on the same pipe run (a surprising insight).

A previous case was quoted concerning a fracture in the weld neck of a pump on an LPG line in an oil refinery, which had caused a fire and had serious cost implications for the operating company. When the pump was stripped out it nearly caused another accident because the two pipes to which it was attached sprung approximately four feet (1.25 metres) axially due to misalignment. The amazing thing is that this pump had been changed a number of times in the past, using an informal procedure for making and breaking the joints – employing two pull lifts and tang spanners – which had been developed to cope with the misalignment! No one had ever thought to cut and re-set the pipes to eliminate the fault. The consequences of this particular piece of poor quality thinking were a fractured pump, a fire, a near-miss incident and a significant loss of profit for the company.

An interesting outcome was that the team, when asked to come up with a definition of the common term 'flogging' (tightening joints by use of a spanner and hammer, a common practice in the petrochemical industry) decided that the best one was 'the uncontrolled tightening of a joint'. It is significant that many more critical joints are now tightened using torque equipment and bolt tensioning than was formerly the case.

12

The communications package

Introduction

In preparing a Turnaround, many thousands of pieces of information are processed into plans and schedules which are created so that the work is carried out in a particular manner. If the objectives are to be met, all involved need to understand what is required of them. Without effective communication the event can go off the rails.

Issues and consequences

A major Turnaround is a complex event, the management of which exposes a number of important issues including, *inter alia*, the following:

- a large volume of work must be done within a very short time;
- the work is performed by a large number of people from many different organizations;
- many different types of work are performed at the same time, in the same place and on different levels of the plant;
- some of the work will be hazardous;
- some of the work will be unfamiliar;
- the onset of emergent work may cause priorities to change, rapidly.

A consequence of this is that the potential for accidents, conflict, error and confusion is greater than normal, so it is vital that everyone involved in the event, or who can influence its outcome, is fully and properly briefed on all its aspects and requirements. The Turnaround manager, assisted by his preparation team, must therefore produce a formal document that is used to brief all personnel and which will ensure that everyone gets the same message. This is the meaning of communication prior to a Turnaround.

Communication

Communication, prior to a Turnaround, is carried out to provide accurate general information, to alert everyone to the rules governing the event, to create a common understanding of requirements and to gain the commitment of the people involved. In a prior briefing session, the following subjects should be covered:

- the purpose of the Turnaround;
- the Turnaround organization;
- key dates and events;
- working patterns;
- the Turnaround work scope;
- contractors;
- cost profile;
- quality plan;
- safety issues;
- facilities;
- the Turnaround objectives.

The general briefing

The general briefing, based on the previously outlined agenda, is delivered to all groups who have a stake in the event. The briefing programme usually starts four weeks before the start of the event. The following groups (each one typically of between 20 and 50 people) must be briefed:

- the Turnaround steering group;
- business and marketing managers;
- the Turnaround control team;
- plant managers, engineers, supervisors and workers;
- local company resource groups (mechanical, electrical etc.);
- all contractors;
- project teams;
- inspectors;
- workshop personnel;
- all support groups (engineering, technology, process etc.);
- all emergency services (both company and community);
- local authorities and other external groups likely to be impacted by the event.

Timing

Obviously, everyone cannot be briefed at once so a briefing programme is arranged. The timings shown below are indicative:

Steering group	–	four weeks before the event
Business management	–	three weeks beforehand
Control team	–	two weeks beforehand
Plant personnel	–	one week beforehand

All other local groups – two to three weeks beforehand
Contractor labour – the day they arrive on site
Casual visitors – require safety briefing only

Delivery

Presentation of the briefing is carried out by a small group of key people, viz:

- The Turnaround manager (who presents the Turnaround issues).
- The plant manager (who presents the plant-based issues).
- The maintenance, or engineering, manager (the engineering issues).
- The safety officer (the safety issues).

In this way the respective elements are delivered by the person best qualified to do so.

Format

The briefing is presented both as a written document for circulation to key personnel and as a series of overhead transparencies to be presented at the briefing sessions. Most plants have a safety video which is shown to anyone visiting the site. If it is used as part of the briefing, it must be borne in mind that it refers to a running plant – which has one specific set of conditions and hazards – whereas the Turnaround will take place on a plant that has been taken off line and opened up. This will present a totally different set of circumstances and hazards. It may well be the case, therefore, that the plant safety video is of little or no use in this context. The decision to use it or not should be taken jointly by the Turnaround and plant managers. Figure 12.1 is a check list of the topics covered by the general briefing.

The major task briefing

Major tasks are those that are large, complex, of long duration or unfamiliar. They will have been planned by a team under the leadership of the preparations co-ordinator and then vetted by the plant engineer in charge of the area in which the work will be executed. With the best will in the world, the engineers responsible may miss a vital element, or remain ignorant of an important piece of information, or of the job method. While being adequate from an engineering point of view, the omission of the detail may pose other problems, not the least of which may be those impacting on safety.

Because these jobs are deemed to be so critical to the success of the Turnaround, they should be subjected to a final validation process, known as the *major task briefing*, during which the area engineers must present the specification and methodology for each selected task to an open forum

Topics to be considered

1. Purpose	2. Organization	3. Key dates and events

– perform statutory inspections
– repair/replace faulty equipment
– investigate process difficulties
– modify existing plant
– install new plant
– introduce new technology
– demolition of redundant plant
	– who is who/where they fit in
	– control team members and roles
	– size and shape of organization
	– different companies/
	   departments involved
	– different teams/their functions
	– different areas/unique features
		– date plant comes off line
		– date plant ready for turnaround
		– date of significant or unusual
		   events
		– date of any interruption to work
		– date of turnaround completion
		– date plant comes back on line

4. Working patterns	5. Work scope	6. Contractors

– normal working day
– normal overtime patterns
– extended overtime requirements
– shift patterns – 1, 2 or 3 shifts
– 8 or 12 hour shifts
– 24 hour coverage
– time off in lieu
– which resources will work
  which overtime or shift patterns
	– capital projects
	– modifications
	– construction work
	– demolition work
	– plant overhaul:
	  • major tasks (critical paths)
	  • small tasks
	  • bulk work
	– unusual or hazardous tasks
		– how many contractors
		– contractor company names
		– which contractor does what
		– specialist contractors
		– interface between contractors
		– interface between
		  contractors and local
		  resources

7. Cost profile	8. Quality plan	9. Safety issues

– overall cost of the event
– cost breakdown by area
– cost breakdown by:
  • management and planning
  • local and contract labour
  • materials
  • tools and equipment
  • services, utilities and
   facilities
– cost of major tasks
– cost of projects and modifications
– any other significant costs
	– approved quality procedure
	  for the turnaround
	– quality auditing (if required)
	– plant quality team and function
	– quality objectives:
	  • only approved material
	   and equipment used
	  • all instruments calibrated
	  • all work completed to spec
	  • critical joints recorded
	  • all joints leak free
		– permit to work system
		– personal protection equipment
		– safety and emergency
		  equipment
		– safety rules and procedures
		– emergency procedures
		– double checking of tagging
		– isolation practice and control
		– working near live systems
		– working with electrical
		  equipment
		– observing barriers and signs
		– personal safety issues
		– personal monitoring
		– general housekeeping
		– hazardous substances
		– safety team and safety cabin
		– safety complaints and
		  suggestions
		– safety initiatives
		– safety targets (drive them home!):
		  • zero accidents, incidents
		  • zero health hazard
		  • zero environmental pollution

10. Facilities	11. Objectives	

(use plot plan to indicate)
– site security and control
– vehicle car parks and
  restrictions
– daily accommodation
– canteens and mess halls
– washrooms and toilets
– emergency facilities
– first aid facilities
– ambulance and hospitals
– out of hours facilities
	– to meet all targets set for:
	  • duration
	  • cost
	  • quality
	  • safety
	– to complete all scheduled
	  work to technical specification
	– to hand the plant back in a
	  condition that will allow it to run
	  leak free till the next shutdown	

Figure 12.1 The general briefing

of managers and engineers, to test their validity. The Turnaround manager arranges the meeting and selects the significant major tasks for each area. The area engineers prepare a detailed presentation on the intended execution of each of the selected tasks.

The forum

The personnel who are invited to the major task briefing are drawn from the following:

• the Turnaround steering group;
• nominated plant and engineering staff;

- safety manager and officers;
- managers of support groups.

The format

The Turnaround manager gives an overview of the major tasks selected and highlights any significant features, such as:

- the ultimate critical path task and the critical path task in each area;
- any special techniques being used for the first time;
- any new technologies being introduced;
- degrees of difficulty of the work to be performed.

Following this introduction, each area engineer in turn presents a detailed description of the selected tasks, using the following format:

- the work to be carried out;
- the planned order of work;
- the techniques and technologies to be used;
- numbers and types of labour to be used;
- any assumptions being made;
- any possible difficulties and the strategy for overcoming them;
- any possible contingencies and allowances for them;
- any known hazards and strategies for eliminating or guarding against them;
- a rescue plan (if required);
- overall cost of the task and specific significant costs.

Members of the forum should be encouraged to ask questions, challenge assumptions and make suggestions during the presentation in order to test its validity. The outcome of this session may be that either the presentation is accepted by the forum as it stands or that modifications are imposed.

The major task briefing should be held about four weeks before the event, to allow time for any proposed modification to be carried out.

Other briefings

Several additional briefings are required prior to the start of the event. They are the responsibility of:

Area engineers — who brief their teams on the area work scope, area objectives, potential problems, roles and responsibilities;

Logistics co-ordinator — who briefs contractors on their specific responsibilities, as well as site discipline and security;

Project managers	–	who brief plant and Turnaround control teams on the scope of their projects, and how they will interact with the Turnaround schedule;
Plant manager	–	who will brief the process, Permit to Work and quality teams on the standards of performance required;
Engineering/maintenance manager	–	who will brief the maintenance and quality teams on the standards of performance required;
Contractor safety officers	–	who will brief their employees on the site safety rules and reinforce the safety messages presented at the general briefing.

Effective communication is vital to the success of the event and the Turnaround manager is responsible for ensuring that this is achieved prior to its start.

13

Executing the Turnaround

Introduction

It is impossible to convey, by means of the written word, the experience of executing a Turnaround, especially a major one. The event is the realization of the plan and its course will be affected by how nearly that plan matches the reality of the event. Tens of thousands of individual activities will have been interlocked in an intricate plan – and it only takes one unforeseen or unforeseeable event to trigger the plan's unravelling.

Managing a Turnaround will provoke the whole range of emotions, from the buzz of excitement which comes from successfully juggling so many balls in the air at one time right through to the pit of despair (usually experienced, in solitude, in the small hours) when a manager finally has to admit, given the current circumstances and constraints, that one or more of the objectives will not be met.

The following processes and routines allow the Turnaround manager to understand what is going on and thereby retain control of the event.

The event

The day finally arrives when all planning and preparation are (hopefully) complete, the materials and equipment are in place and the resources have been organized and briefed. The business of shutting the plant down and executing the Turnaround begins. Although many thousands of varied activities will take place during the event, the Turnaround manager should determine that from his perspective as a work manager there are, in fact, only two types of work that have to be controlled, routine and unexpected.

The routine

Work that has been planned, scheduled, and resourced and, where necessary, appropriate allowance has been made for contingencies. The emphasis here should be on exercising control over this work, by means of the Turnaround planning and control system, to meet or beat the schedule and budget.

The unexpected

A Turnaround is a complex event during which problems may arise which have been neither predicted nor expected – no matter how good the planning and preparation. They could be caused by such things as changes of intent, accidents, industrial conflict and so on. The aim is to be prepared – to react speedily and effectively to the situation, minimizing any negative impact. If the incident is not properly controlled the routine can rapidly become the unexpected and the unexpected may become the catastrophic.

Shutting the plant down

Shutting down is controlled by the plant manager and his personnel. It should be carried out in conformance with the shutdown network plan, which, as part of the overall Turnaround schedule, will have been formulated or updated by the plant team during the preparation phase. In many situations, especially those in which the plant is being shut down for the first time, or by an inexperienced team, the logic, timing and duration of events will be based only on judgement and on whatever experience is available. Reality may force the plant team to deviate from the plan; if this happens it can have two effects:

(i)   it alters the logic of the shutdown network and affects the timings and durations of activities;
(ii) it affects the start of the mechanical duration.

The plant manager should inform the Turnaround manager so that the effect of the deviation on the existing schedule can be rapidly calculated, and the schedule amended if necessary. Failure to do so may render the existing schedule useless as a control tool.

Prior to the start of the shutdown phase all necessary tools, equipment, materials and isolation plates are identified and laid out on the plant so that they will be available when required. The Turnaround manager supports the plant manager and may supply resources to carry out the following supporting tasks:

- transporting shutdown equipment around site (hoses, drain valves etc.);
- fitting and removing isolation plates;
- positioning and connecting water cooling and washing equipment;
- fitting shutdown valves and pumps;
- breaking and remaking joints to the plant team's instructions;
- cleaning up spillage of fluids and other substances;
- carrying out any other task which progresses the shutdown;
- the plant team 'cools' the plant down and decontaminates it before handing it over to the Turnaround manager. A plant may be handed over in total or one system at a time.

Some of the typical faults that cause the shutdown to over-run and intrude into the time allowed for the mechanical duration are:

Problems with physical items – items which are necessary to shut the plant down (tools, plates etc.) are either not there when they are required or are found to be unsuitable.

Duration over-run – activities take longer than planned. This may be due to the planned time being unrealistic (it may have just been a guess) or the process used to perform the activity being ineffective (e.g. four tonnes of steam being used when 10 was required).

Emissions and spillage – product or service fluids are emitted or spilled and have to be cleaned up before work can proceed.

Inexperience – personnel employed to shut the plant down not having the necessary experience to do so in a timely and effective manner.

Disorganization – the shutdown team leader loses control of the work due to bad planning, poor briefing or ineffective communication.

Effects of the shift system

On shutdowns which take more than one shift to complete, time can be lost if the shifts have different approaches to shutting the plant down or the incoming shift insists on re-checking some of the work carried out by the outgoing one.

Tight management of the shutdown on a twenty-four hour basis by experienced shift managers who have agreed the logic of the shutdown, stick to it and perform an effective handover at the start of each shift will go a long way to eliminating these problems.

The routine (see Figure 13.1)

The Turnaround manager's routine

A Turnaround can typically last anything from one to six weeks. In order to remain in control, the Turnaround manager must set up a daily routine which will allow him to deal with all critical items and stay in touch with what is going on. It must include personal contact with key people in the organization on a regular basis. It also does no harm (and often much good) if he talks to the workmen on the job, often gaining insights into the situation which he would otherwise miss.

To a great extent, the Turnaround manager must rely upon other people to feed him the information he needs to make decisions and take actions but it would be a foolish manager who allowed that reliance to become a dependency. He must be able to verify that the information fed is in accord with the overall progress and performance of the work.

The most effective way for a turnaround manager to control the rate of progress on an event is by means of a daily routine.

The daily routine helps the manager keep his finger on the pulse of the event and apply pressure when and where it is needed.

Actions

- Check previous 24 hours progress with the planning co-ordinator
- Check cost control and forecast with the cost co-ordinator
- Visit safety cabin and check on any safety issue
- Visit the stores and check on material delivery and issue problems
- Visit the workshops (if feasible) and check on work progress
- Tour the site to check on safety and housekeeping – talk to people!
- Regularly take part in safety inspections and spot checks
- Visit permit-to-work office and check on any problem
- Visit the quality team and discuss any quality issues
- Vet overtime requests and approve/amend/reject

Meet with the plant managers and:

- resolve any current technical problems
- discuss and approve/reject requests for emergent work
- formulate strategies to keep the event on programme
- discuss and resolve any industrial relations problems
- discuss and resolve any interface problems
- define the consequences of any change of intent

Chair the daily turnaround progress meeting:

- safety officer reports on safety issues and initiatives
- area engineers report on area work progress and issues
- project managers report on project progress and issues
- plant manager reports on any plant related issues
- maintenance manager reports on any engineering issue
- quality team leader reports on quality issues
- cost co-ordinator reports on expenditure and cost issues
- chairman sums up/makes decisions/delegates actions

Write a daily progress and safety report, and issue it

Figure 13.1 Turnaround manager's daily routine

A typical daily routine for a Turnaround manager will now be outlined. It may not always be possible, nor even desirable, to carry out the activities in the order shown but it is important that they are carried out with daily regularity. Everyone else on the Turnaround takes their cue from the Turnaround manager and if they can discern a regular pattern they will be more comfortable.

The process used to fulfil the purpose of the routine is:

- examine the situation to gain understanding;
- decide if there is a problem and, if there is, define it;

- expose the root cause of the problem;
- create a solution for the problem;
- take action or delegate action to implement the solution;
- record the action taken;
- monitor the effectiveness of the solution.

The process must become automatic because solutions to the many problems which arise during an event must be created on the run. There is little time for deliberation and everyone else is looking to the Turnaround manager to keep things on track – he is probably the only person who ever sees the whole picture. Anyone who manages a Turnaround should get plenty of opportunity to perfect this skill.

The Turnaround manager's daily routine should include, but not be limited to, the following procedures.

Night shift progress

The Turnaround manager should check, first thing in the morning, the progress and performance for the previous night shift. On the Turnaround schedule, for the purposes of controlling resources and durations, work done on night shift is treated in the same way as work done during the day, but it has to be remembered that at night there is usually only a skeleton control crew on duty and most of the key personnel are off duty.

Normally, night shift is used to progress the network and any work which has fallen behind schedule – the most crucial work on the Turnaround – and yet it is often the least managed activity. It only takes one thing to go wrong, an incorrect briefing, issue of the wrong materials, an equipment breakdown, to halt work. Getting it started again can be difficult because the normal services and personnel are not available. Even call-out takes time and sometimes fails due to bad communication.

Critical path hours (which may represent 30 to 40 per cent of the total hours allotted to the job) that are lost on night shift can never be recovered. Night shift work needs to be controlled by a strong and resourceful manager and should not be used to blood inexperienced engineers. Night shift planning, preparation, resourcing and briefing needs to be of the highest standard.

Control of work

The Turnaround control team will have generated a number of documents to control work and the manager must keep up to date with current progress by visiting the planning office (or cabin) daily to get an update on the progress of the event. He should examine and analyse documents, including, among others:

The Turnaround schedule – comprising the shutdown network linked to the mechanical duration schedule and the start-up network. This is updated daily. It shows the overall progress in every area and indicates whether a job has gone critical (i.e. fallen behind planned schedule).

Look ahead schedules – abstracted from the Turnaround schedule and show the work needing to be done over the next few days. They are especially useful to the supervisors who are executing the work on site.

Run-down graphs and S-curves (see Figures 13.2 and 13.3) – showing actual and projected performance against planned performance. The variable plotted can be cash expenditure, or completed programme-hours, or expended man-hours, etc.

Work control sheets – list batches of similar jobs (vessel inspections, scaffold builds etc.) and show the key stages of each job where, as part of the single point responsibility principle discussed in Chapter 7, signatures are required to indicate that that particular stage has been completed.

These are all visual aids which allow the manager to see, at a glance, the state of progress overall and on all the individual areas.

The validity of the schedules and control documents depends upon their accurate updating. If this is ignored a false picture of progress can be created. There is a practice among some supervisors of 'keeping some

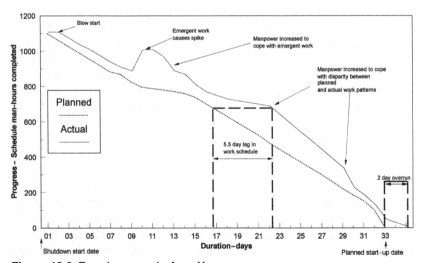

Figure 13.2 Run-down graph: Area X

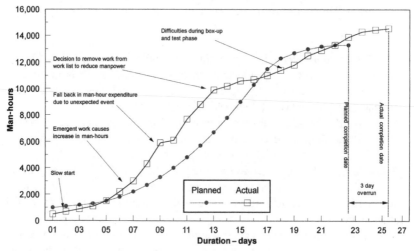

Figure 13.3 Example of an S-curve: Expended man-hours

hours in the back pocket'. This means that they fail to reveal all of the progress that has been made – especially if they are ahead of programme on a particular job – because they know from experience that problems will occur as they go through the event and if they are held up they can feed some of the 'back pocket' hours into the record and appear to be working on schedule. This practice distorts, propelling into the critical zone work which has already been completed but not reported. The schedule may even indicate to the Turnaround manager that he is behind schedule when he is not – prompting him to take measures to respond to a situation which does not exist. The manager who does not have his ear to the ground may be stunned by the apparently huge amount of work which is completed on the final day as supervisors dump all of their remaining back pocket hours into the schedule. The work is completed on time maybe, but the manager may already have expended time, resources and money to recover from the fictitious backlog.

The other side of the coin is represented by serious problems with performance or progress which are not reported by the area teams – lulling the manager into a false sense of security. The Turnaround manager may believe he is on programme when, in fact, he is well behind. Again, but with more serious consequences, he does not find out until it is too late to do anything about it. The Turnaround manager must know the true situation on the event if he is to exercise control and this knowledge only comes from constantly monitoring and questioning the updating information.

Critical path job (CPJ)

Although duration is associated closely with control of work, its importance merits a paragraph or two to itself. At all times, the Turnaround manager must be aware of a number of factors which have a direct bearing on the critical path (which will determine the actual duration of the Turnaround).

Most other types of problem on an event may be resolved within the planned end date by the judicious use of overtime, extra resources etc. This is unfortunately not true of a job which lies on the critical path; programme time lost there can normally only be recovered by taking work out of the CPJ. However, this is only very rarely possible at any time during the event and virtually impossible late on in the event.

To ensure that the state of the critical path is known, there are several questions which the Turnaround manager should ask on a daily basis, viz:

1. Which job is currently the CPJ and what type of work does it entail?
2. Is the CPJ on, ahead of, or behind schedule? If behind – why, and what can be done about it?
3. Are there any known issues that will cause problems on the CPJ?
4. Is there anything that will cause a job not currently on the critical path to become the CPJ, if so, can anything be done to avoid this?

The Turnaround manager and his control team should devote some time every day to an analysis of the current situation on the critical path, in order to keep it under control as far as possible. The worst case scenario – which reflects, as nothing else can do, the uncertainty at the heart of a Turnaround – is the emergence of a significant piece of extra work on the critical path job (e.g. damage is found, or a pressure test fails on a welded butt and weld repairs are required), which adds to the duration of the event, and nothing can be done about it. In this case, the Turnaround manager must involve the steering group who must analyse the effects and publish a new duration for the event.

Expenditure

The cost co-ordinator should produce a daily report detailing current expenditure (actual and committed) and AFC (anticipated final cost) for each of the designated areas of the Turnaround as well as for the overall total. This information allows the Turnaround manager to analyse expenditure patterns and pin-point activities which are overspending. This can occur unexpectedly in straightforward activities such as scaffolding and lagging, rather than in the more highly technical activities where it might be anticipated, but which are often more tightly planned and controlled. The manager is able to expose root causes and propose remedial action. Also, the report is backed up by more detailed information which can be used as necessary. This is especially important if a large amount of emergent work is driving expenditure upwards.

Copies of the report should be sent to the members of the steering group to keep them in touch with the situation because, if the AFC increases beyond the forecasted budget (based on the final estimate) the two main responses are to find more money to cover the extra work or to prune the work list – either way the policy team must be involved. Prior warning is preferable to an unpleasant surprise when it is too late to respond (e.g. finding out that the event has overspent after it is finished).

Safety

The Turnaround manager should visit the safety cabin and check on any current safety issues. *Every day* – there is always news on safety and it's usually bad. It is one thing to formulate a safe system of work during the relatively calm days of the preparation phase when everyone is thinking rationally. It is quite another to enforce it on hundreds (in some cases thousands) of people who are working under time pressure and (as has been proved so often in the past) will act irrationally and thereby put the health and lives of themselves and others at risk in order to progress the job.

Even more irrational is the almost obstinate refusal of some people to wear the protective equipment they are issued with and obey the rules which have been set up purely for their protection. Housekeeping is a very mundane subject which is often ignored and yet bad housekeeping generates unsafe conditions which are a major contributory factor to accidents.

The Turnaround manager has ultimate responsibility for the safety performance of the event. He must exercise constant vigilance and monitor the safety situation regularly in order to ensure that the safe system of work is operated to the full.

Logistics

The logistics teams should be visited and the current situation and any anticipated difficulties noted. The current disposition of tools, equipment, materials, proprietary items, consumables, services, utilities, accommodation and facilities must be understood and any problems, requiring the input of the manager, resolved.

Items for special consideration should be:

- non-delivery of materials
- disposal of toxic or effluent substances;
- violations of site rules (such as use of restricted areas for storage etc.);
- utilization of cranage, heavy plant and vehicles;
- any issue regarding accommodation or facilities which is likely to cause industrial relations problems (a shortage of toilet paper can be as troublesome as a shortage of materials!).

Workshops

The Turnaround manager (or his nominee) should regularly visit workshops to check on progress. Nowadays, most workshops are off-site and it is imperative that the old adage 'out of site, out of mind' does not apply. If progress is not monitored, not only can work fall behind schedule but the workshop may be working to the wrong priority. The items being worked on must be delivered back to site before or on their due dates. A check should also be made on the state of items being delivered to the workshop to ensure that any required on-site decontamination has been carried out. If marshals are being employed to organize the movement of bulk items between site and workshops they should be included in the regular workshop visits.

Site safety

Monitoring is a vital component of the safe system of work. Using the unsafe acts/unsafe conditions routine, the manager should tour the site daily to check on safety and housekeeping. This must include talking to as many people as time allows and constantly pressing home the safety message. The Turnaround manager should also regularly take part in formal site safety inspections and spot checks.

Permits to work

Permit to work issuers must be visited and any problems discussed. Permits are, on many overhauls, a constant source of frustration and wasted time. Given that the permit system has been set up to cater for the unique conditions of the event, permits should be available when required. If they are not, the root causes of the delays must be exposed and dealt with. With the increasing use of contractors, the client management are often faced with large claims for delays due to late permits. There is no easy answer to this. Only real co-operation between all parties concerned, constant monitoring of the situation and a willingness to act to improve when required will keep wasted time to a minimum.

Quality

There should be regular visits to the quality team to discuss any problems. The team will normally have a monitoring role in specific areas (such as the making of joints) throughout the greater part of the event and they are normally the plant representatives who sign off the contractor's work on its completion. If time pressure is resulting in shoddy workmanship (at any level) the penalty will have to be paid during start-up when such workmanship will be exposed in the most expensive way – the plant will not start up or function properly. A moment's thoughtlessness can have grave consequences. Again, if there are quality problems, the root causes must be exposed and eliminated.

Working patterns

The Turnaround manager should carry out a daily check on working patterns to check that they actually match the needs of the event. Shift patterns (especially those of the night shift) must be regularly checked against the amount of work which is being achieved. The daily overtime requests from Area Engineers must be analysed and approved or rejected. A daily check should be made on productivity and manpower utilization.

Co-ordination

The plant and engineering/maintenance managers should be met regularly to:

- resolve any technical problems;
- discuss, and approve or reject, requests for extra or additional work;
- formulate strategies to keep the event on schedule;
- resolve any industrial relations problems;
- resolve any interface conflicts;
- discuss and define the consequences of any change of intent.

Control (see Figure 13.4)

A daily control meeting should be chaired, with the following agenda:

1. Safety officer's report.
2. Area engineers' reports.
3. Project managers' reports.
4. Plant managers' report.
5. Engineering/maintenance manager's report.
6. Quality team's report.
7. Turnaround manager's summing up and actions.

When the meeting is over, the Turnaround manager writes his daily report and issues it. The report should contain, but not be limited to, the following:

- a brief overview of the progress and performance of the Turnaround highlighting any major problems or achievements;
- a safety statement highlighting good and bad practices, brief reports on accidents, the daily and cumulative safety statistics and the following day's safety theme;
- if applicable, a statement on current costs and manpower levels.

The report should be restricted to one sheet of A4 paper and circulated as widely as possible.

Information and actions for consideration

1. Safety officer's report

- details of accidents/incidents in last 24 hours
- details of accident/incident trends
- findings of daily site inspection and spot check
- summary of site safety level and details of any particular safety concerns
- recommendations for safety improvements
- details of any safety initiatives or awards
- recommendation for tomorrow's safety theme

3. Project manager's report

- progress to date on project including any technical problems and solutions
- progress on any 'break-ins'
- any 'bad fit' problems due to poor design
- any hold ups or shortages of manpower, materials, equipment or services
- any conflicts with other areas of work
- whether the area is on schedule or behind – and, if behind, the strategy for getting back on target
- assessment of unavoidable over-run, how many hours or days, and why it is unavoidable

5. Maintenance manager's report

- any concerns on turnaround progress
- any engineering concerns
- any quality performance concerns
- any upcoming engineering problems
- any questions on turnaround engineering work being done

7. Area engineers' reports

- progress on major tasks including any technical problems and solutions
- tasks completed, boxed up and handed back and percentage completion of other major tasks
- progress on small tasks and bulkwork
- any hold-ups or shortages of manpower, materials, equipment or services
- any conflicts with other areas of work
- whether the area is on schedule or behind – and, if behind, the strategy for getting back on target
- assessment of unavoidable over-run, how many hours or days, and why it is unavoidable

2. Turnaround manager's routine

- open the meeting and control through the chair
- ask for and note reports in pre-set order
- ask specific questions to clarify points
- do not allow detailed discussion of issues at this meeting – convene separate discussions
- sum up general progress on key indicators
- voice any concerns on trends or specific issues
- make executive decisions and inform the meeting of them, their requirements and consequences
- delegate specific actions to particular people
- delegate responsibilities to convene further discussions on key issues outside of the control meeting
- announce next day's quality initiative
- announce next day's safety slogan
- make any other announcements
- state, and ask for, any other business
- close the meeting

4. Quality team leader's report

- quality trends in the last 24 hours
- any specific quality problems
- any recurring quality problems
- recommendations for quality improvement

6. Plant manager's report

- current ability of permit to work issuers to issue permits on time and strategy to eliminate any delays
- any handover quality issues
- general view of on-site performance
- general view of on-site housekeeping
- upcoming on-site problems
- warning of any system coming back on line early
- warning of any process activity that could impact progress or safety

Figure 13.4 The daily control meeting

The daily Turnaround programme (see Figure 13.5)

As well as his own daily routine, the Turnaround manager should regularize the activities of the event's other key personnel by issuing a daily Turnaround programme. Unlike his own routine, which can have a certain amount of flexibility from day to day, the timings of the activities on the daily programme once set, are fixed. This is because the outcomes of many of them are fed into subsequent activities. Because of the large amount of work that has to be carried out on a Turnaround, having a daily programme is essential. The more 'in control' of the routine work the team is, the more time is available for dealing with the unexpected.

Time	Activity	Personnel
0700	Report on overnight activities to area co-ordinators	night shift engineer, co-ordinators
0715	Prepare permits to work and issue Marshal materials and equipment and collect permits – review task sheets	permit issuers, co-ordinators, supervisors
0730	Brief area team on day's work	area engineer
0800	Labour reports for work and are 'checked' on the job	supervisors all trades
0805	Work teams briefed on requirements	supervisors
0815	Day's work commences	all personnel
1015	Morning coffee break	all personnel
1130	Formulate overtime requests for that evening	co-ordinators, supervisors
1145	Approve/modify overtime request	manager
1300	Lunch break (thirty minutes)	all personnel
1430	Update of work progress against plan for all areas of the turnaround	area planners, supervisors
1500	Site logistics meeting to programme following day's crane lifts and other rigging requirements	logistics officer, crane co-ordinator, rigging foremen
1530	Area team meetings to programme last twenty-four hours progress and safety issues	area engineer, area planners
1545	Feedback on day's work to work teams	supervisors
1600	End of normal day's work. Men are 'checked off' the job	supervisors, all trades
1600	Turnaround control meeting to report to manager on progress, raise issues and take action	manager/engineers safety officer, plant management
1630	Overtime teams commence work	supervisors + men
1930	Handover briefing for nightshift team's work programme	supervisors, night shift engineer

Figure 13.5 Example of a daily Turnaround programme

The unexpected (see Figure 13.6)

The purpose of planning and preparation is to reduce what is unknown to an absolute minimum, and to routinize everything that is known. As discussed in Chapter 1, there are several uncertainties that lie at the heart of a Turnaround, because it involves a plant that is worn or damaged, and to an extent that is unknown. Also, it is prone to the inherent vagaries of all human activities – we disagree, make mistakes, get tired, and change our minds without warning.

Because the Turnaround is a task-centred event one uncertainty which can have a significant effect is emergent work, which falls into two categories: extra and additional, which were defined in an earlier chapter. For the purposes of this section, however, extra work is defined as that which is generated by an existing task (such as repairing a crack found during an inspection) and additional work as tasks that were not part of the original plan but were inserted or requested during the Turnaround, either because they were missed off of the original work list or they have been generated by a change of intent or a by a previously sound piece of equipment failing during the event.

Outline procedure for handling technical problems

1. A fault is discovered during the execution of work or inspection
2. The area engineer analyses the fault to define the extent and consequences of the problem
3. Area engineer contacts turnaround, plant and maintenance managers and appropriate technical experts and convenes a meeting
4. Area engineer presents the problem to management team and appropriate technical experts and suggests possible solutions
5. If necessary, a turnaround policy team meeting is convened and the team informed of the problem and possible solutions
6. The team decides upon a course of action and responsibilities are delegated as follows
7. Technical experts produce necessary instructions, drawings and documentation needed to carry out work
8. Site logistics officer procures material and equipment
9. Area engineer and planning officer calculate the effect of the extra work on the turnaround schedule
10. Turnaround manager and quantity surveyor calculate costs
11. Area engineer gathers all necessary information and forms it into a proposal defining work requirements, costs, timing and duration of the work
12. Management team reconvenes to approve, rework or reject the proposal
    - if approved, the work is carried out on the due date
    - if reworked, the proposal is then approved
    - if rejected, either operations 7–11 are repeated until approval is obtained or the work is not done
13. When the work is approved, the planning co-ordinator amends the turnaround schedule to include the new work

Figure 13.6 The unexpected

The Turnaround manager, in consultation with the plant and engineering/ maintenance managers, must approve all emergent work. In the case of extra work, allowance for such a contingency may have been built into the plan and budget. This will not be so with additional work unless, in the light of experience with previous Turnarounds, a lump sum has been put into the budget to cover costs. However this alone will not address any negative impact on the schedule; emergent work (especially of the extra variety) may extend the duration of the Turnaround. This effect is more critical when it happens near the end of the Turnaround because there is then little or no time left to recover.

Emergent work may change the critical path from one task (or area) to another and it is then vitally important to closely monitor its effect. Resources may otherwise be concentrated on a task which is not currently the critical path, while the actual critical path task is robbed of the resources it so badly needs.

If the effect of emergent work is significant, the Turnaround manager must have the authority to reconvene the steering group, at any time during the event, to discuss the impact of the work and make whatever changes to the policy or objectives that this necessitates. The types of change that emergent work, if it must be done, can force upon the steering group are:

- increasing the budget to cover extra expenditure;
- declaring a longer duration;
- altering the logic of the schedule;

- eliminating other, previously planned, work from the Turnaround;
- altering resource levels, shift patterns or overtime levels;
- stopping work on a less critical area of the event and switching resources.

Whatever the changes, they will have a negative impact on routine and they are bound to cause frustration among those members of the group most affected. The Turnaround manager must move to minimize this and motivate his workforce. Also, the costs of emergent work must be closely monitored and kept separate from the Turnaround budget.

Once the emergent work has been discovered, there must be a routine put in place to handle it as professionally as possible (see Figure 13.7). This should process the work through initial identification, registration and endorsement (or rejection). Once endorsed it is technically specified by the appropriate level of planning and then submitted for final approval. If this is given the work is funded, resourced and integrated into the work schedule before being handed over to the person responsible for carrying it out. If the work is rejected at the endorsement or the approval stage this must be recorded to prove that the work has been rejected only after due consideration, and not forgotten or rejected out of hand. Even in this there must be single point responsibility.

The power of routines

By means of the routines for planned work and for coping with unexpected work when it occurs, the Turnaround manager strives to keep control of the event. Very occasionally, a situation will still arise which completely swamps the system and makes the normal routines inadequate – a serious accident that causes a walk out, a defect so large that it will keep the plant off line for weeks or months. For the most part though, effective routines will handle almost any situation.

Starting up the plant (see Figure 13.8)

A point is reached in each area of the Turnaround at which most of the tasks have been completed and the Turnaround manager and plant team together agree that the area may be handed back to the latter for start-up. This is a critical transition phase and if it is not properly controlled, time, money and effort can be wasted.

Although it may seem to the casual observer that the start-up of the plant is simply the reverse of the shutdown, there are some significant elements which differentiate the two events, viz:

- There are usually many tests – pressure tests, system tests, loop checks and trip and alarm tests – which have to be carried out during the start-up period which make it more complex than shutdown. Also, any one of these tests may fail and need re-working, extending the event's duration. This means that, in effect, there may be two or more work programmes running in parallel, either of which can delay the other.

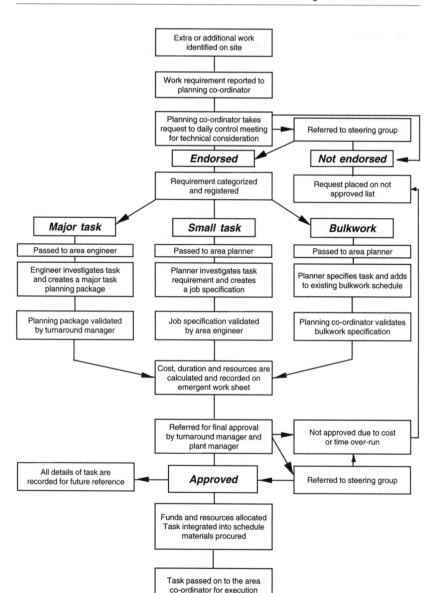

Figure 13.7 Emergent workflow diagram

- The start-up comes at the end of the event when most people have been working long hours over an extended period of time. Tired people make more mistakes.
- Unlike the shutdown when the plant is being cooled and de-pressurized, and fluids are been extracted, during the start-up the plant is being heated and pressurized, and fluids are being introduced, increasing the hazards.

- Have all items on the work list been completed?
- Has the area team carried out a quality check e.g.:
  - Does new flanged pipework have correct gasket fitted?
  - Have supports been fitted to new pipework?
  - Have new stress-relieved valves been repacked?
  - Have all joints got correct gaskets fitted?
  - Have instrument modifications been checked and tested?
  - Have all new valves been repacked/rejointed?
- What post start-up work remains, has it been listed and agreed by the start-up engineer/co-ordinator?
- Have all outstanding clearances been signed off?
- Have all control sheets been completed i.e. site mods, critical joints, vessel inspections RVs, CVs?
- Have all isolations been checked to ensure all slip plates removed?
- Do any joints in the area require 'hot flogging' – has it been arranged?
- Are the de-staging lists complete e.g. hot stagings, pre start-up staging, post start-up stagings?
- Is any special lighting required during start-up?
- Has any instrument damage been identified and corrected?
- Has all equipment been removed where possible. If not, list what remains e.g. weather protection, scrap, welding sets etc.?
- Do the plant operators understand the implications of the site modifications carried out?
- Can operators gain access to valves etc.?
- Have the shutting down defects been identified and corrected?
- Has all extra work been executed and any start-up implications communicated?
- Are there any critical CVs/RVs which need overhauling to be available during start-up?
- Are the insulation lists up to date, pre start-up/post start-up?
- Are there any control valves in your area which need close attention during the start-up?
- Have all electrical checks been carried out?

Figure 13.8 Handover check list: area team to start-up team

- There is a possibility that, on shutdown, critical path time lost may be recovered and the event put back on programme. On start-up, however, such time lost cannot be recovered, it simple extends the duration.
- There have been thousands of individual activities performed on plant equipment during the shutdown and the mechanical duration. Any one of them may have been incorrectly performed and cause a fault to emerge during start-up.
- An item of equipment that was not on the Turnaround work list may fail on start-up. This is especially true of pumps.
- An isolation plate inadvertently left in position during start-up can have consequences ranging from frustrating to catastrophic.

Depending upon the type of plant, the start-up process can last anything from a few hours to many days. Whatever the length of time, the handover must be done effectively. The plant manager and his team control the start-up in the same way in which they control the shutdown and, as then, the Turnaround manager provides resources to support the plant team.

The Turnaround manager cancels the daily control meeting and convenes in its place a daily start-up meeting, the emphasis of which is on getting the plant back on line as safely and quickly as possible. The following personnel are normally involved in the meeting:

- Plant manager – Chairman
- Turnaround manager – Technical adviser
- Turnaround engineer – Work organizer
- Turnaround co-ordinator – Punch list organizer
- Turnaround planner – Work planner
- Logistics co-ordinator – Material procurement
- Control instrument engineer – Trip and alarm systems
- Safety representative – Control of hazards
- Any other nominated person

The team meets at regular intervals during the day to review progress and update work requirements. The agenda for the meeting is as follows (see Figure 13.9):

The start-up meeting agenda in detail

1. The punch list
- specify the current list of outstanding work
- review the previous day's punch list
- strike off all completed work
- investigate causes why any planned work was not completed in the previous 24 hours
- re-define priority tasks/changes of priority
- define required completion dates/times
- highlight any hold up or shortage for each job
- highlight any job which cannot be completed
- create strategies to overcome problems
- assign resources to jobs
- responsibilities for getting work done

2. Extra/additional work
- define any extra or additional work that has been generated by the start-up itself
- highlight any work which must be brought to the attention of the turnaround steering group
- add the work to the punch list
- set priorities/completion dates for the work
- if necessary, reorganize the punch list to accommodate the new work
- specify manpower, material, equipment and services requirements
- if required carry out a task hazard assessment
- (plant personnel) is there any potential extra or additional work likely to be generated in the future, based on experience of past start-ups

3. Procurement
- record any deliveries in the last 24 hours
- define any existing procurement problems
- delayed/long delivery/unavailability
- delivered items rejected/need for concession or other action
- assign responsibility for procuring manpower, materials, equipment and services for newly generated work
- assign cost centres for purchase of resources

4. Start-up key dates/problems
- specify current start-up logic
- list key dates for individual system start-ups
- specify any change in priority from previous day's meeting
- define problems that are delaying start-ups
- define any potential problem which might delay start-up
- create strategies to overcome problems
- specify any unavoidable over-run of start-up programme
- if necessary, inform the turnaround policy team of expected over-run

5. Site demobilization
- demobilization of turnaround team members
- local manpower being sent off site
- contractors leaving site today
- equipment being taken off hire
- services being demobilized
- accommodation and facilities being sent off site
- return to stores/vendors all unused items and materials
- washing and cleaning programme to eliminate effluent and other unwanted substances
- housekeeping programme to return site to its original clean and safe state

6. General points
- final closing of punch list and any exceptions
- closing out or cancellation of all work orders and permits to work
- date and time of final site inspection by plant, maintenance and turnaround managers
- date and time of post mortem debrief
- promised date for turnaround final report
- promised date for turnaround quality manual (if required)
- date/venue of turnaround topping out party

Figure 13.9 Starting the plant up

1. Punch list of all outstanding work.
2. Additional work generated by the start-up.
3. Procurement of items generated by the start-up.
4. Start-up activities/key dates/problems.
5. Site clean-up.
6. Any other issues.

The start-up team needs to be closely co-ordinated, responding quickly to the ever-changing situation which characterizes the start-up. The start-up control meeting continues until the plant is back on line and productive and the plant manager is satisfied that all technical work has been completed. At that point the start-up team is de-mobilized and plant personnel return to their normal duties.

14

Terminating the Turnaround

Introduction

It is the Turnaround manager's responsibility to arrange all the activities that will return the plant to an acceptable condition, at least as good as it was before the event. He must ensure that the handover is properly concluded and that all traces of the event are removed, ensuring that the environment of the plant is at least as good as it was before shutdown. He must then arrange a final site inspection and obtain a handover certificate from the plant manager, indicating that the Turnaround is over. Finally he must de-brief the key personnel who worked on the Turnaround and, from the information gained, write a final report.

Demobilizing the site

Most of the work of demobilizing the site is arranged by the logistics co-ordinator, but it is still the responsibility of the Turnaround manager to ensure that it is carried out in a proficient and proper manner. Demobilizing refers to the removal from site of all people, goods and services temporarily housed there for the duration of the event. This includes, *inter alia*:

All manpower resources, both local and contractor – The manpower demobilization should be phased to match the declining workload. The rule is to demobilize manpower at the earliest opportunity.

All surplus materials, unused proprietary items and unused consumables – These will either be returned to the stores, sold, recycled or dumped. If returned to stores, all material certificates or fitness-for-purpose certificates must be returned also. (It is ironic that some of the modern spares management systems make it difficult, if not impossible, to return surplus items to the stores.)

All tools, equipment, cranage and vehicles – Once again, hired equipment must be taken off-hire at the earliest possible date.

Turnaround control cabins and fitments – This includes cabins, furniture, computers, telephones, faxes, photocopiers etc.

All temporary accommodation – Stores, mess rooms, toilets etc.

All temporary structures – Scaffold, platforms etc. The plant team may request that some temporary structures be left in position to allow them to perform future work. In this case the costs should be transferred to the plant budget.

Any temporary work stations – Water washing bays, quarantine compounds etc. All should be removed and thoroughly cleaned (no effluent should be left behind).

Unwanted elements – Redundant plant, spent catalyst, debris, rubbish, dirt, spillage and effluent should be safely removed.

Demobilization includes a thorough cleaning of the plant area and the removal of any other item or substance temporarily installed. In addition, any damaged painted surfaces should be patched or painted over and all insulation re-installed (except where the plant manager has requested that it be left off).

The final inspection and handover

When the Turnaround manager is satisfied that the site has been fully demobilized and cleaned, he arranges a final site inspection with the plant manager, accompanied by the logistics co-ordinator and any person nominated by the plant manager (it is also a good practice to take a small team of cleaners who can clean up any area of the plant which does not meet the plant manager's standards). They visit every area of the plant and inspect to ensure that:

- all agreed work has been completed;
- all traces of the Turnaround have been removed;
- the plant is clean and tidy;
- any damage done during the Turnaround has been repaired.

The plant manager should identify anything that is not satisfactory. The logistics co-ordinator records it on a punch list. In addition, the plant manager may request certain omissions, e.g. that particular insulation be left off, or scaffolding be left in position because there is further work, unconnected with the Turnaround, to be done on line. These omissions need to be recorded because the cost of completing the work will be transferred to the plant manager's budget.

The logistics co-ordinator arranges for any outstanding work to be carried out and, if necessary, for the Turnaround manager and the plant manager to then re-inspect the plant. When the plant manager is satisfied with the state of the plant, he signs a handover certificate which indicates in a formal manner that the Turnaround has been completed.

Recording performance – Turnaround metrics

When all of the work has been completed and the plant is back on line, it is time to take stock of actual performance against planned performance. Measurable data is collected from many sources and recorded on a metrics sheet.

Figure 14.1 shows an example of a comprehensive performance review record sheet, covering safety, cost, duration, hours, efficiency and quality and other measurements against the objectives for a Turnaround. A space is also provided for recording the critical factors that contributed to actual performance.

Once all of this data is available we can start to measure, in overall terms, how successful the Turnaround was.

The other important point about formally gathering and recording the metrics for the Turnaround is that it provides us with crucial information on past performance when we initiate future Turnarounds. It is a vital part of the package of information that allows us to link Turnarounds that may be separated by a number of years into links in a continuous chain.

While it may be true that each individual Turnaround is a 'project' in its own right, when taken together the performance of successive Turnarounds over a period of years becomes a 'process' – more a business process – and, as has been demonstrated by systems science time and again, the lifeblood of effective processes is the feedback of information on performance. The importance of this simple routine cannot be overstressed.

Shutdown..........................Area.............................Date.........

Safety				Cost			Duration			
Class	Plant	Client	Contractor	Allocation	Estimate	Actual		Plan	Actual	Comment
							Shutdown			
				Late work	Extra work	Additional	Schedule			
							Start-up			

Work-scope hours			Efficiency				Quality of work			
At closure	Late work	Total plan	Class	Mech	Fab	Inst/Elec	Welds	Press test	Joints	Handbacks	SU leaks
			Productivity								Normal
Extra work	Additional	Total	Utilization				Failed	Failed	Failed	Rejected	Critical
			Plan factor								

Cost breakdown

Management	Planning	Inspection	Materials	Hired equipt.	Client lab	Contract lab	Specials	Waste and rework

Critical factors contributing to actual performance

1	
2	
3	
4	
5	
6	
7	
8	
9	

Figure 14.1 Performance review record sheet

The post-mortem debrief

The Turnaround has been completed. The personnel who performed it have all returned to their normal jobs. The plant is back on line. The site has been cleared and the handover certificate has been signed. However, that is not the end of the story. All of the planning and preparation that went into the event, all of the briefing and control mechanisms, were aimed at achieving one goal, performing the Turnaround within the pre-set targets for safety, quality, duration and budget. The Turnaround may or may not have achieved one or more of these. Some tasks will have gone well and others will have been a disappointment. Targets may have been missed for a number of reasons which only hindsight can determine. For instance:

- the targets may have been unrealistic;
- the policies imposed (by the steering group) may have been counter-productive;
- the planning and preparation may have been inadequate;
- communication may have been sub-standard;
- contractors may have been incapable of meeting their promised targets;
- emergent work may have overwhelmed the ability to meet the targets;
- unexpected events may have overwhelmed the ability to meet the targets;
- concentrating on achieving one target may have caused others to be missed.

If the organization is to be improved, it must analyse the recently completed Turnaround – to measure actual against planned performance; to ascertain the root causes of success and failure so that features that caused good performance may be reinforced, and those that caused bad eliminated. This is the business of the post-mortem debrief.

A word of caution here. There is always the risk that a debrief can turn into a witch hunt. If an aspect of the event has gone badly and management are more interested in finding someone to blame for poor performance, rather than finding the root causes of the fault, the whole affair may degenerate into a blood letting. Everyone becomes defensive and nothing is learned. Heads rolling (whether the victim was actually the guilty party or not) may satisfy certain individual managers within the company, but will not promote consolidation of the wealth of information. The negative approach does the company no good and it will remain ignorant of the causes of poor performance, especially when the whole process will have to be repeated in two, three or four years' time, with the risk that the same mistakes will be made again.

It is vital that the Turnaround manager sets the trend by ensuring everyone understands that the debrief, when it is held, will be a learning event, not a search for the guilty. He should agree a date for the debrief with the plant manager and any key personnel required to attend. The meeting should be chaired by him and subjects that might well be being dealt with could include:

- The scope of work on the original schedule. The amount of extra and additional work that was carried out and the reasons why it was necessary.
- Planned versus actual duration and the reasons for any differences between them.
- Planned versus actual man-hours, resources, shift patterns, overtime levels and the reasons for any differences between them.
- Targeted quality performance versus actual performance and the reasons for any failures to meet the targets.
- The actual safety record – highlighting any accidents or incidents and the findings of the subsequent inquiries.
- General impressions on overall performance.
- Lessons to be learned from the analysis.
- Recommendations for future improvement.
- Closing comments by plant manager and Turnaround manager.

NB The above list is not intended to be exhaustive.

The number of debriefing meetings required will depend upon the number of people involved.

But the debrief meetings themselves are not enough. The company needs to have a process of capturing the learning from the event and formulating the recommendations for improvement into an action plan. Figure 14.2 shows the process for turning the lessons learned from the debrief meetings into an action plan for improvement. The product of the action plan is fed back to the Turnaround team, fed upward to the business managers and, most important of all, fed forward into the planning for the next Turnaround. In this way we connect each of the individual Turnarounds (which may be separated by years) into one continuous process.

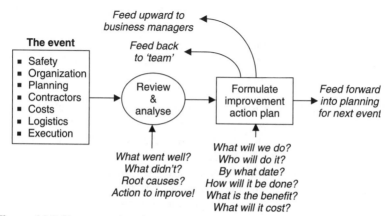

Figure 14.2 Turnaround performance review

There are two important points to make about this process. The first is that the debrief meetings should be held as soon after the event as possible when people's memories of the event are fresh, and the second is that if there is no action plan to improve then the debrief meetings are a waste of time. This may seem obvious but, in my experience, there are companies where the 'lessons learned' report from Turnarounds lie gathering dust on the shelf while the company continues to make the same mistakes year after year.

The final Turnaround report

The Turnaround manager now has all the available information concerning the event. His final task is to organize it into a Turnaround report. Typically, the general topics covered would be:

- Turnaround policy
- The work scope
- The preparation phase
- Planning
- The organization
- Control of work
- Contractor performance
- Safety
- Quality
- Logistics
- Communications
- Recommendations

Figure 14.3 is a detailed checklist of the requirements for each topic.

Once the report is completed the Turnaround manager issues copies to all key personnel. It is the link between one Turnaround and the next, which ensures (if it is used properly) that the lessons of the past will contribute to an improved performance in the future. It is the responsibility of all key personnel to examine the report, analyse its message and recommendations and add to the sum of knowledge on which the company depends for its success and survival.

The Turnaround report is the final activity in a process which has lasted many months: a process which, when driven by a rational methodology, goes a long way towards ensuring that the prior objectives of the Turnaround are met to the satisfaction of the management of the company.

The next action is to decide the date for the first meeting of the policy team for the next Turnaround, and the first document which should be considered at that meeting is the report from the previous Turnaround.

The final report should contain the following information

1. Turnaround policy	2. The work scope	3. The preparation phase
Breakdown of actual performance measured against initial objectives – actual vs planned duration – actual vs planned cost – actual vs planned work scope including extra/additional work – actual vs planned man-hours including overtime and shifts – actual vs planned accidents/incidents	– a description of each major task – a list of minor tasks – numbers of bulkwork jobs – a description of each project – a list of statutory inspections – any registered inspections – instrument and electrical work including trip and alarm tests – a list of defects	– turnaround management and planning hours expended – pre-shutdown work – long delivery items – systems set up for reception storage and issue of materials – a copy of the approved plot plan – special problems encountered
4. Planning	**5. The organization**	**6. Control of work**
– planning system used and any comments on its performance – comments on number and quality of task sheets and lists – comments on number and quality of job logic networks – plant validation of task sheets – any special problems	– a copy of the organization sheet – comments on the size and shape of the organization – comments on the effectiveness of the organization – comments on work management teams – comments on plant and specialist support	– computer printed schedules – check lists – control sheets – 'planning boards' – 'S' curves – daily meetings – effectiveness of reporting
7. Contractor performance	**8. Safety**	**9. Quality**
List individual contractors and for each, detail actual performance against planned performance for: – duration of work – quantity of work done – quality of work – overall cost (including extras) – safety Score each contractor (out of 100) and recommend if they should be employed again	– summarize safety performance – list each accident/incident in detail with causes and inquiry recommendations – comment on housekeeping – comment on wearing of PPE – compare contractor performance with local resource performance – comment on effectiveness of the permit to work system – comment on the effectiveness of the emergency system	– define quality team function – comment on quality team performance – highlight any major quality non-conformances – comment on contractor quality performance – comment on any specific quality initiatives
10. Logistics	**11. Communications**	**12. Recommendations**
– outline the logistics plan – comment on actual facilities vs planned facilities – comment on performance of the logistics team – highlight any specific difficulties	– describe briefing programme – comment on turnaround team internal communication – comment on plant/turnaround team communication – comment on turnaround team/contractor communication – comment on communication of daily safety initiatives – any other general comments	Write a conclusion on the overall effectiveness of the turnaround. Recommend improvements in any of the eleven categories covered by the report – based upon: – any safety inquiry – any technical inquiry – the post mortem debrief – personal experience – the experience of anyone who volunteered recommendations

Figure 14.3 The final turnaround report

Conclusion

Success in Turnarounds, like success in any other walk of life, depends upon a number of basic requirements. These requirements have been detailed in this part of the book.

Safety – have we taken all of the precautions necessary to provide an effective safe system of work for the people involved in the Turnaround?

Organization – are the people competent to do the work, are they properly organized and are they led effectively?

Planning – has the work been specified, controlled, planned and scheduled within the required duration, and cost?

Contractors – are they 'fit for purpose', effectively briefed and properly integrated into the organization?

Costs – do we know how much the event is going to cost and do we have a plan for controlling expenditure.

Logistics – can we guarantee that the right thing will be in the right place at the right time? And

Execution – do we have the routines and the control measures to allow us to carry out the work in a logical and rational manner? Finally, do we have the strategy and systems to complete all of this safely and to the desired level of quality?

If the honest answer to all of these questions is YES, then the Turnaround has the best possible chance of succeeding.

Asking these and other critical questions was the process that led to the creation of Part 2 of this book. The questions were refined and formed into what has come to be known as the Model of Excellence for Turnarounds. Part 2 introduces the model and deals with a number of case studies where the model was used in the real world.

Part 2

The Model of Excellence

15

Real world application

Introduction

This part of the book differs from Part 1 in two significant ways.

First, Part 1 dealt with the *process* – a series of activities – for planning and executing a Turnaround, whereas Part 2 introduces a *Model of Excellence* for Turnarounds that was distilled from applying the process, exploring that application and discovering the basic principles that underlie excellent performance.

Second, the process defined in Part 1 was presented as an ideal process and was free of context whereas the Model of Excellence recognizes the difference in operating context from company to company and is concerned as much with *how* an activity can be carried out within a certain context as with the activity itself.

For example, one basic principle of Turnarounds is that they are complex engineering events that need to be accurately planned. The 'ideal' approach cited in Part 1 is that the planning should be done by a team of experienced planners (in my opinion still the best way to do it), but to put this into a specific context, what if the company does not employ planners?

The answer to this question will be determined by what resources the company does have available and their attitude to seeking help from external sources.

This understanding generates a whole spectrum of possible approaches to planning a Turnaround. At one end of this spectrum, I have been involved in a Turnaround that used a carbon copy of the approach recommended in Part 1 of this book and at the other end of the spectrum I have been involved in a Turnaround that was planned (and managed) by production technicians, because that was the best way to do it given the company's circumstances. Both Turnarounds were successful.

I guess the best way to express this is that 'what' you need to do (basic principles) is fixed, and 'how' you do it (methodology) is flexible. It is a company's ability to align the fixed principles and flexible methodology that will determine its effectiveness in performing Turnarounds.

Of course, the acid test is that, having decided upon an approach within a given context, we must then answer the question 'is it adequate?' and this is achieved by the auditing element of the Model of Excellence.

The conceptual models explored in Part 2 of this book are based on the real world approach of designing portions of Turnaround methodologies based on the basic principles of the Model of Excellence applied to a company's current resources and operating context.

Shifting the paradigm

The theory behind Part 1 of this book is that there is a rational, logical process for doing Turnarounds. As far as it goes it is a sound theory. However, the process will only work effectively in the real world if it has a supportive environment. That means an environment in which, for example, the stakeholders in the Turnaround (maintenance, production, business etc.) co-operate with one another.

Another even more fundamental requirement is that those involved in the Turnaround (at all levels) actually understand what is going on. I originally assumed that this would be the case. More fool me!

I managed a Turnaround for a company who had engaged a new production manager two months before the event start date. During the Turnaround he often bore a vaguely puzzled, worried expression and he seemed to be struggling to come to terms with what was going on.

A few days after the event was completed he wondered into my office, sat down and said. 'Could you tell me what just happened?' I asked him what he meant and he replied, 'That Turnaround we just did. It's the first one I have been involved in and for most of the time I hadn't a clue what was going on. Everything seemed to be happening haphazardly. . . .'

As he spoke he was waving his arms in circles in front of him as if trying to give some kind of shape to what he was thinking. I asked him what the gestures represented. He thought for a long moment and then replied, 'It's like a big black cloud with flashes of everything happening all at once.'

I tore off a sheet of flip chart paper, put it on the desk in front of him and asked him to draw the cloud. He spent the best part of an hour drawing it. Figure 15.1 is a 'cleaned-up' version of the drawing. It nicely represents the confusion of someone who is unfamiliar with shutdowns. It is the first slide I show when I am teaching courses on Turnaround management and when it comes on the screen I watch the faces of the participants closely. There are always some of them who smile ruefully and nod in agreement as if to indicate 'I know that feeling.'

This incident, more than anything else, was the catalyst for the shift in my Turnaround paradigm. I had understood 'the process' but it was obvious that other people didn't. The critical question for me was 'Why is that?'

After a lot of analysis I determined that Turnarounds were unfamiliar to many people involved in them for one or any combination of the following main reasons:

1. Turnarounds are considered to be 'technical' events not requiring much input from senior management, so there is no Turnaround business strategy.
2. The management has no real understanding of the equipment in its plant so there is often uncertainty about what constitutes an effective work scope.
3. Turnarounds may be infrequent events involving different people each time they are done, so they become unfamiliar events.
4. The company is incapable of learning from past experience.

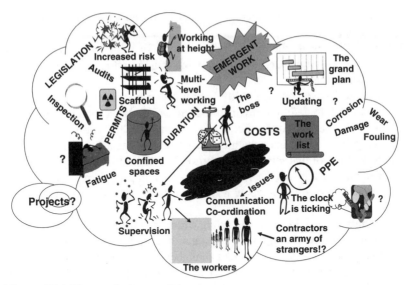

Figure 15.1 The confusion model

5. The company does not train its personnel in Turnaround planning and execution. Many companies are unaware that such training exists.

6. The company does not possess written standards, procedures or instructions for performing Turnarounds but depends on the experience of individuals.

The next thing was to determine what could be done to rectify this situation. Re-examining the 'process' in Part 1 of this book, what struck me was that it took 179 pages of script and diagrams to explain the process, principally because Turnarounds are complex events and the process requires that much definition.

The process is an effective tool, it has been proven to be effective time and again but it is an approach suited to a company where the above listed short-comings had been eliminated. It is an 'ideal' process that assumes the company has completed the journey from innocence to excellence.

To be robust, what was needed was a model that would assist companies to travel that road, from innocence to excellence and had a sharper focus that provided a more immediate way of representing not merely the steps in the process but the basic principles that underlay the process.

More than that, if the model was to be powerful enough to lead companies from innocence to excellence it had to be utilitarian and be capable of satisfying a number of stringent conditions, as follows:

1. **It had to be comprehensive enough to cover all critical Turnaround elements and yet simple to understand and communicate to others**.
    The current outline model covers 13 critical Turnaround elements that are broken down into 82 'areas for consideration' and yet is represented

on a single sheet. The detailed model is further broken down into 321 critical activities.

2. **It had to provide the means for a company to audit or assess its current Turnaround approach and be capable of generating recommendations for improvement.**

   To date I have conducted 30 + audits using a more detailed version of the model. In addition, my mature MSc and diploma students in UK and Australia have used the model to perform some 150 audits in their own companies. Their overwhelming response has been that the model is a powerful tool that cuts to the heart of Turnaround issues, easily exposes weaknesses and allows generation of real improvement. Many of the companies involved adopted recommendations proposed by their mature students.

3. **It had to provide a basis from which a company can develop an effective methodology for planning and managing Turnarounds.**

   Using the model, I have assisted a number of companies to either improve their existing methodology or create a completely new methodology. In addition my students have created a number of Turnaround methodologies in their own companies.

4. **In line with point 3, it had to form the basis from which a company could produce a set of written standards, procedures and instructions for Turnarounds.**

   I have used the model so far as a basis for writing a full set of Turnaround procedures for four major companies and specific procedures for other companies. In addition, a number of my students have produced procedures for their companies based on the model.

5. **It had to be capable of forming the basis of a training programme for Turnaround management.**

   I have used the model to create a Turnaround management training course that has been taught in various forms in a number of countries around the world.

6. **It had to be auditable and flexible enough to incorporate improvements generated by the actual use of the model in the real world.**

   Each time the model is used it is 'assessed' for effectiveness. The model is now in its ninth version. The original model had six main elements that covered the technical aspects. The latest (but not final) version has 13 elements. The change has come about through using the model in the real world and adding or changing critical elements based on actual experience.

The model

The result of the distillation of basic principles from the original process and the learning from real world application was a model based on best practice

SAFE SYSTEM OF WORK	MANAGEMENT OVERVIEW	TOTAL QUALITY APPROACH
Objectives – Safe Systems – Assessment Compliance – Monitoring	**Constantly Test and Improve Turnaround Strategy**	**Standards – Q Assurance – Q Control Compliance – Monitoring**

ORGANIZATION	PLANNING	CONTRACTS	COSTS	LOGISTICS	EXECUTION	TERMINATION
Steering Group	Planning Philosophy	Contract Policy	Pricing Framework	Logistics Approach	Management Control	Start-up Control
Event Manager	Project Definition	Contractor Commitment	Ballpark Estimate	Procurement Control	Briefing Programme	Critical Activities
Designed Organization	Work Scope Definition	Critical Elements	Estimate Refinement	Infrastructure Design	Human Factors	Start-up Routine
Planning & Prep Team	Work List control	Incentive Schemes	Contract pricing	Supply Control	Planned Daily Routines	Plant Start-up
Technical Team Input	Detailed Planning	Subcontract Policy	Contract pricing	Bulkwork Organization	Plant Shutdown	Post Start-up Monitoring
Production Team Input	Schedule Optimization	Contractor Mobilization	Cost Control	Offsite Services	Scheduled Work	Post-mortem Debrief
Resource Levels	Pre-event Criteria	Performance Measurement	Cost Monitoring	Cranes & Transport	Emergent Work	Close-out Meeting
Contractor Integration	Schedule Updating	Contractor Demobilization	Closing-out Accounts	Site Clearance	Equipment Hand-back	Action to Improve
Risk Management	Risk Management	Risk Management	Risk Management	Risk Management	Risk Management	Prepare for Next Event

COMMUNICATION STRATEGY	IMPROVE via REPEATABILITY	AUDIT
Communications Policy Activities – Spread	**Capture & Re-use Best Practice Eliminate Bad Practice**	**Audit Protocol: Strategy – Preparation – Execution**

Figure 15.2 Outline Model of Excellence

from a large number of Turnarounds carried out in many companies in different parts of the world. For this reason it was named the Model of Excellence for Turnarounds.

Figure 15.2 shows the Model of Excellence in outline. There are 13 crucial elements and each has a guiding principle in order to focus attention on the critical requirements of a Turnaround. They are:

Element   Guiding principle

1. Organization   Professional 'One team' design

This is concerned with the people who will plan, prepare and execute the Turnaround and is based on two fundamental tenets: (1) that the organization be designed to suit the requirements of the job – rather than patched together from available people, and (2) that all of the disparate organizations involved are integrated (so far as is possible for an ad hoc team) into 'one team' so that everyone has a common objective.

2. Planning   Rational 'task-centred' work control

The task is at the centre of the Turnaround and task requirements generate every one of the Turnaround requirements (money, manpower, methods, materials, machines). Therefore the work list needs to be as close to an ideal work list as possible and must be stringently controlled. This is to ensure the best possible inputs to the planning system to optimize detailed plans and schedules.

3. Contractors   Focused 'win–win' relationship

The fundamental requirement is to reach a position where the client is getting value for money and the contractor is making a reasonable profit. This implies mutual respect and close co-operation based on the client taking the lead in negotiating with contractors so that each party understands the needs of the other.

4. Costs   Aggressive 'value for money' strategy

The Turnaround manager is given stewardship of the company's money and it is incumbent upon him to ensure that the company gets value for the money it spends and that, as far as is possible, all activities that waste money are aggressively exposed and eliminated.

5. Logistics   Dynamic 'right–right–right' approach

At its most basic level, logistics concerns getting the right thing in the right place at the right time. In order to achieve this a dynamic approach is required where there are people in the organization who are made responsible for the procurement, reception and distribution of all materials, equipment, services and facilities required to perform a Turnaround.

6. Execution   Streamlined 'results-oriented' approach

Taking all actions necessary to find the shortest, safest and simplest path from the beginning to the end of the event and keeping everyone focused on the end result. This involves a great deal of co-ordination. There is a saying that 'no plan ever survives the first day of execution'. Effective execution is about responding to reality and taking what actions are necessary to hold to the spirit of the plan even when the substance changes.

7. Termination   Serving the wider system

Once the scheduled (and emergent) work is completed we need to ensure two critical things happen. The first is that the plant is handed back to the production or operations team in a fit condition so that the return to full production can be accomplished by the swiftest and smoothest means. The second is that we must capture all of the learning from the event and feed it into our 'Improvement via Repeatability' programme.

8. Safe system of work   Safety first, last and always

Turnarounds are inherently hazardous undertakings and it is incumbent upon management to provide the highest level of safety to safeguard the health and safety of people and the protection of the environment. Only through a managed safety programme can this be accomplished.

9. Total quality approach   Getting it right first time

To be successful the shutdown process requires inputs that are fit for purpose. In computing terms there is an acronym 'GIGO' which means garbage in > garbage out. If we do not procure the right people, materials, equipment and services then we cannot expect to achieve the objectives of the Turnaround.

In the model, quality has two aspects – quality assurance which involves actions taken before the event to ensure inputs are fit for purpose, and quality control which involves actions taken after the performance of the task to ensure it conforms to standards.

10. Communication strategy   Say what you mean to say

Effective communication is absolutely crucial to the success of the Turnaround. We promote good communications by establishing a strategy that allows us to specify what information is required, who will receive it, by what means it will be transmitted and when it will be transmitted.

11. Audit Program   Inspect what you expect

We learn most by auditing what we do (when we are doing it), analysing the findings of the audit and making recommendations for improvement as we go along. Audits are carried out to examine the proposed management strategy for the Turnaround, the systems to be employed on the Turnaround and people's compliance with the systems provided.

12. Improvement via   Focus on continuous improvement
repeatability

Closely allied to audits is the review stage of the Turnaround where we collect, collate and analyse the comments of those who were involved in the event. When we understand what went well and what didn't, and the root causes for inadequate performance and specify action to improve, we can record our findings and recommendations and use them as the basis for an improvement programme that will be fed forward into future Turnarounds, ensuring that we improve continuously.

13. Management overview   Integrated business strategy

Management is the only level that can have an overall view of Turnarounds and it is only by understanding the events in a business context and continuously challenging how we perform them that the company can reap the full benefits from all of the effort that is put into planning and managing a Turnaround.

The model is a generic template that specifies the critical areas that should be *considered* when planning a Turnaround. Note the use of the word 'considered', this is deliberate and crucial to the model.

The reasoning behind the Model of Excellence is that it represents best practice and if all of the principles specified are followed effectively the result should be a successful Turnaround. However, in application, the model also recognizes that the individual approach taken by any given company will be based on its unique situation and its culture. The actual working model is far more extensive and detailed than the outline model shown in this book but even with the outline it is possible to address the areas of activity and assess their effectiveness.

So, when the management uses the generic template as a guide to develop the company's approach to Turnarounds they consider each of the elements of the template. They may choose to accept or reject any element of the model. That is their prerogative. However, there needs to be a sound reason for rejecting an element.

But, there is always another side to the coin. Having worked with the model for a number of years I have found that the amount by which a company deviates from the principles of the model will determine the reduction of its chances of success.

The model as an auditing tool

As mentioned earlier a large number of audits have been successfully carried out using the Model of Excellence.

Remember, the principles and critical activities of the Model of Excellence have been formulated by the experience of hundreds of people working on live Turnarounds. It is not a theoretical construct. When you measure your approach against the Model of Excellence you are benchmarking your company against current best practice.

What follows is a protocol for what I define as a brief audit. The brief audit findings may generate a more in-depth audit.

The simplest way to use the model is to ask questions similar to the following for each critical activity:

- Does this exist? or Do we do this?
- What process do we use to do it?
- How well do we perform each step in the process?
- What difficulties do we experience when doing this?
- What results do we get from doing this?

Brief audit protocol

Using the Model of Excellence as a template assess the company's approach to Turnarounds by comparing it with the 'ideal' approach in the model.

Step 1

(a) List the people interviewed.
(b) List the documents interrogated.

(c) Produce a matrix, based on the Model of Excellence, which allocates a numerical score, based on your findings, to each of the critical activities using the following criteria:

N/A This activity is not performed and is not, in my opinion, required
0 This activity is not performed but should be
1 This activity is performed but could be significantly improved
2 This activity is performed and requires only moderate improvement
3 This activity is used effectively and requires no improvement

Calculate the subtotal for each element and an overall total for all thirteen elements.

Step 2

Write a report that highlights the following factors:

(a) An executive summary that gives a brief outline of the report.
(b) The matrix complete with scores.
(c) The strongest elements of the company's approach.
(d) The weakest elements of the company's approach.
(e) The root causes that underlie the weaknesses.

Step 3

Formulate a list of recommendations for improving the company's Turnaround approach (this should include a feasibility/cost/benefit statement for each of the recommendations).

Step 4

Submit the report to the relevant authority for approval to implement the improvement recommendations.

Step 5

When approval has been obtained, formulate an action plan that defines:

- The specific action in as much detail as necessary.
- The name of the person who will be responsible for the action.
- The date by which the action is to be completed.
- An estimated cost for the action.
- The estimated benefit likely to result from the action.

Step 6

Evaluate the Model of Excellence in terms of it usefulness as an auditing tool and suggest any improvements that would make it more effective for future use.

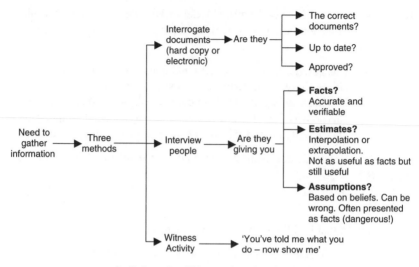

Auditor's credo – If it's not written down it never happened

Figure 15.3 Information gathering model

In addition to the model I have found it useful to create an 'information gathering model' simply to remind people not to take anything at face value when auditing.

Figure 15.3 shows the information gathering model.

Careful examination of the information that is gathered and, wherever possible, cross-checking it and witnessing the actual activity go a long way to eliminating spurious information. Finally, if you want to prove that you have done something you need to record it formally.

I always place the caveat on this type of audit that it must result in action to improve otherwise the whole exercise is a waste of time, money and effort.

Into the real world

Proving the model's effectiveness required real world application. In the following chapters I have detailed six short case studies based on work done with a number of companies. Each case study looks at a particular element of the Model of Excellence and explains how it was used to generate a real world solution to practical problems and issues.

A final word before we proceed

One of the routines used in the Model of Excellence is called 'Logical levels of intervention'. In order to effect change we must know at what level of the organization to intervene and ensure that the intervention is of the right type.

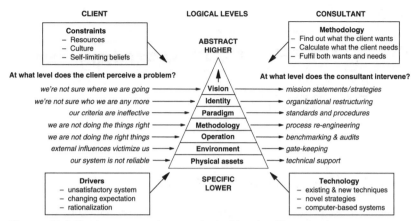

Figure 15.4 Business consultancy – logical levels of intervention

Figure 15.4 shows a diagram I have developed over the years to aid me in thinking this issue through.

The sort of questions it raises and seeks to answer are:

- How does the client see the problem and what does he want done about it?
- How does the consultant see the problem and what needs to be done about it?
- Do the client's 'wants' and the consultant's 'needs' coincide?
- At what level should intervention be made to satisfy the wants and needs?
- Does the client understand the drivers and constraints involved in the desire to change?

By posing these and similar questions at the very start of the improvement programme, the consultant (whether he is internal to the client's company or external) can get a clear picture of the context of the problem as well as its content, what type of problem it is, what level it is on and what is the best way to proceed.

I find this approach reduces the risk of a false start to the improvement project and acts as a signpost to guide subsequent action.

16

Case study 1– Designing an organization

Step 1 – The presenting problem

Client No. 1 had performed two Turnarounds over the previous four years with unsatisfactory outcomes. The events had taken longer to complete and cost more than the planned targets. The feeling among the client's personnel was that although the event had been well planned, they had 'lost the plot' on the day due mainly to what they termed as a lack of co-ordination. In the client's view, performance had always been satisfactory until four years ago.

Step 2 – The assessment

I carried out an assessment audit six months before the programmed start date of the Turnaround using the Model of Excellence and discovered that, traditionally, the teams that executed the Turnarounds had not been formally planned but had evolved over a number of years based on available personnel and depending upon the wealth of experience that those people had gained (but never recorded) over the years.

The root of the company's problem seemed to be that there had been a number of downsizing events over the last ten years (with the last and most significant one five years previously) and a good deal of experience had been lost. In recent years they had been struggling to resource Turnarounds and felt that they had 'lost their way'.

There was no one left in the company who understood Turnarounds or could specify the functions required to perform one effectively. Although there had been a sea change in the nature and form of the client's organization the management had not responded to the change and were still trying to execute Turnarounds as they had always done. This was clearly inappropriate.

For example, the engineer who had 'managed' the Turnarounds in the past took early retirement and, as there had been no succession plan for the position because it was not seen as a 'permanent' position, coupled with the fact that there were no written procedures for Turnarounds, the resultant loss of corporate knowledge had severely impaired the company's ability to manage Turnarounds effectively.

The company had merely appointed another engineer to take over the next Turnaround. However, this was a young engineer who had been involved in only one Turnaround previously and did not have much project management skill. After his 'failure' two years previously he had been replaced by another engineer – who promptly travelled the same painful route.

It was clear that a new strategy was required for the Turnaround organization.

Step 3 – The response

In line with the theme of the Model of Excellence, my recommendation was that the company return to basic principles and design a Turnaround execution organization using a simple form of gap analysis that asks the following four simple but powerful questions.

1. What do we need to ensure success in the future?
2. What do we have now?
3. What is the gap between what we have and what we need?
4. What actions do we have to take to close the gap?

I have found this method to be much more effective than the normal 'this is the problem – how do we solve it' approach because it starts with a target solution rather than trying to move from the problem situation to some unknown solution. I have found it especially useful in complex and 'messy' situations where there is no simple solution to the problem.

I worked with the engineering improvement manager to draw up an eight point procedure for designing a Turnaround organization.

Two discrete kinds of knowledge were employed to ensure an effective organization. The first was knowledge of Turnaround systems and organization, which I provided, and the second was local knowledge of the plant and the available people provided by the engineering manager and other company personnel who became involved in the design of the organization.

Designing an organization is an iterative process – this means that you need to cycle through the process a number of times before you get the optimum solution.

Figure 16.1 shows the procedure that was used in this particular case.

Step 4 – A framework for action to improve

The process to improve the organization was an involved one and for the purposes of clarity I will deal with only part of the organization design.

Traditionally the engineer who managed the Turnaround would organize the planning and package the work for about 16 'supervisors' (actually maintenance fitters with the exception of two supervisors) who would all report directly to him (see Figure 16.2).

1. Analyse the known work list and the geographic layout of the plant and design an initial outline organization that serves the needs of this particular Turnaround.
2. Design an 'empty box' organogram based on the question 'What specific roles do we need to perform the Turnaround?'. At this stage, each box in the organogram requires only a job title.
3. List the people available to fulfil the roles in the organogram and assess the competence of each individual based on aptitude and experience.
4. Match the people to the functions and fill in the names of those we intend to use in each of the boxes. This results in one of three conditions:

   (a) The name of an individual in the box assessed to be competent to carry out the required role.
   (b) The name of an individual in the box who is not fully competent but is the best available choice.
   (c) No name in the box because there is no one available within the company to carry out the required role.

5. Compare 'what we have' (actual organogram) with 'what we need' and expose any shortcomings.
6. Take action to compensate for each shortcoming (e.g. providing training and practice for specific skills or recruiting competent personnel from external sources).
7. Write a set of roles and responsibilities based on the specific requirement of each function. (Note: general roles and responsibilities are often too vague or do not sufficiently cover particular requirements.)
8. Brief (and if necessary train) everyone on their roles and responsibilities to ensure they understand them. This is best done by face-to-face briefing. Briefing by e-mail is not sufficient.

Figure 16.1 Designing the organization

This seemed to have worked reasonably well with the original engineer – probably through long years of practice working with experienced supervisors. But having 16 supervisors reporting to him overwhelmed the younger engineer and within a few days of starting, the Turnaround lapsed into fire fighting mode, from which it never recovered.

What was needed was an organization with a chain of command that could cope with the co-ordination issues generated by the Turnaround. To this end,

Figure 16.2 Flat organization with no definition

Figure 16.3 3-Tiered organization with definition

the role of full-time Turnaround manager was created. This role would be responsible for the planning and execution phases of the Turnaround.

The plant was then divided into three geographical areas and the role of 'area Co-ordinator' created to manage each area. They would report to the Turnaround manager and would have a number of supervisors reporting to them.

As one of the areas was primarily instrument/electrical work it was decided that the co-ordination for this area should be an I/E engineer.

If you compare Figure 16.2 with 16.3, there are two main differences. The first is that the supervision in Figure 16.2 has no definition – meaning that every supervisor looked after his own work package and had no influence (and normally no interest) in what the other supervisors are doing. This led to a very difficult co-ordination situation with 16 separate elements needing to be co-ordinated by one person, the Turnaround manager.

In contrast, the structure in Figure 16.3 decreases complexity by reducing the number of direct reports to the Turnaround manager from 16 to three and delegates responsibility for each of the three geographical areas to one person – the area co-ordinator. They, in turn, have three supervisors reporting to them. This is a much more manageable structure.

The second difference is that, although there is one more layer in the organization and this may buck the current trend of 'flatter is better', there are actually fewer people in this fragment of the organization – 13 as against 17 in the original organization. This makes the organization more cost effective.

Step 4 – Defining outline responsibilities

In order to define what was needed for the various positions we next drew up a set of outline responsibilities and the links between that were required to ensure that the organization operated effectively. Figure 16.4 shows the conceptual model of the responsibilities.

In order for the organization to work it was necessary to specify the outline responsibilities. But as well as this we had to ensure that each level was properly linked to the levels above and below. Each level had a responsibility to: report to the level above; delegate to the level below; monitor the

Figure 16.4 Execution organization hierarchy

performance of the level below and resolve any issues beyond the scope of the level below.

If these links are in place and operating effectively then the organization is a coherent entity, but if even one of the links is broken it can seriously affect the organization's ability to function.

Step 5 – Matching people to roles

Now that we knew what we needed and had defined the various roles in outline we had to match the available people to the organogram in a first run exercise. The result is as shown in Figure 16.5.

Figure 16.5 Execution organization – availability

Shortcomings

*1. The engineer selected for the Turnaround manager's job was experienced in maintenance but had never managed a Turnaround before. However, he did have some project management experience. Overall, this was less than ideal and was considered to be a shortcoming.
*2. The supervisor selected to run Area 1 had a lot of experience as a supervisor on Turnarounds but had not been tested at the higher level of co-ordinator. This was a developmental opportunity for the person considered to be the best of the supervisors but it was also seen as a shortcoming.
*3. There was no one available in the maintenance department to fill the role of Area 3 Co-ordinator. Most of the available personnel lacked experience at the supervisory level let alone the co-ordinator level.
*4. There were only eight people eligible to act as supervisors, one short of the ideal number.

Strengths

1. The I/E engineer was very experienced and had been involved in Turnarounds for a number of years and had also managed an I/E upgrading project the previous year. He was regarded as a strength.
2. The eight people eligible for the role of supervisor were assessed to be capable of carrying out the role. This was also seen as a strength.

To compensate for the above shortcomings an action plan was drawn up as follows:

1. I was engaged by the client to directly coach the Turnaround manager and assist him in developing the systems for, and the planning and preparation of, the Turnaround until such times as he felt confident to do it on his own.
2. The candidate for Area 1 Co-ordinator was sent on a public course on Turnaround management and then spent a month at one of the company's other plants acting as an assistant to one of the area engineers on a large Turnaround.
3. The Area 3 Co-ordinator's role was filled by an engineer from the project department who had extensive experience on managing projects on Turnarounds. He also attended the public course on Turnaround management. He was seconded to the maintenance department for three months.

   An added bonus was that the project engineer took the responsibility of briefing the supervisors during the run-up to the Turnaround on their roles and responsibilities.

4.   There was no one available in the client's organization to fill the ninth supervisor role. We negotiated with the contractor who normally supplied mechanical fitting resource to provide us with a person who had experience in supervising Turnarounds. In actual fact the client was forced to hire two contractor supervisors because one of the originally nominated supervisors became long-term ill and was not available for the Turnaround.

During the following five months a number of 'get-together' sessions were organized in order to build team spirit. All concerns were openly aired and a number of brainstorming exercises were carried out to resolve specific issues.

One month before the event, team members were gathered together at a session to risk assess the organization. The engineering improvement manager and Turnaround manager led the session. The outcome – an aggregate level of confidence among team members that the organization would be 90 per cent effective.

The outcome

The Turnaround was subsequently completed by the designed organization within the budget and with one day over-run on duration. The over-run was due to emergent work and it was estimated that, had it not been for the emergent work, the Turnaround would have finished half a day ahead of plan.

For many, the most dramatic change was the absence of fire fighting and the calm and organized manner in which the work was carried out. One supervisor who had been involved in previous Turnarounds was heard to remark *'This doesn't feel like a Turnaround at all. No buzz, no excitement. If it goes on like this it could get boring!'*

17

Case study 2 – Work list control

Step 1 – The presenting problem

Client No. 2 had a chronic planning problem. The planning team could never get their planning and schedule completed before the start of the Turnaround and as a consequence, the Turnaround (especially the first few days) was disorganized. They felt that they always allowed enough time to plan and were at a loss to understand why they always seemed to start the Turnaround with an unfinished plan. They guessed that this was costing them money but could not quantify it.

Step 2 – The assessment

Using the Model of Excellence I carried out an assessment on the planning approach three months before the event. The first thing I found was that they indeed allowed enough time for planning – six months for a medium sized event was more than sufficient. They also had two very experienced planners whose technical planning skills were of the highest quality.

I examined the quality of specification from the job requesters and found it variable but mainly of reasonable quality. Those were the pluses. I also found two large minuses.

When I investigated the number of people authorized to request work I found that it was everybody! Anyone on the plant could input a job request onto the Turnaround work list and it was accepted without question. As a result, the work list was very messy.

I carried out an impromptu challenge exercise on the Turnaround work list with the planners and one of the production supervisors. Among other faults on the 312 job work list, we found the following:

- *26 of the job requests had been duplicated by people on different shifts*;
- *17 of the jobs were deemed to be unnecessary*;
- *1 request was for a change-out on a pump that no longer existed*.

One of the unnecessary jobs requested an internal inspection on a large tube heat exchanger. When the job requester was asked why the job had been requested he replied: '*Well, it's been four years since we looked in there and I thought it was about time we did.'* When asked if there were any indication that there was anything wrong with the heat exchanger he replied: *'None, but just to be on the safe side I would like to look inside.'* Finally he was asked if any faults were picked up the last time it was inspected and he replied: *'No, it was as clean as a whistle – but you can never be too careful.'*

Because of its size and location within the plant, the fact that a large crane and a specialized bundle puller would be required and an extensive scaffolding platform would need to be built, the estimated cost for this job was £21,000. The team considered that this was taking preventive maintenance too far and the job was cancelled. A rough estimate of the cost of the 17 cancelled jobs gave us a figure of approximately £110,000.

The problem with the duplicated jobs was that they were often planned before the duplication was discovered, wasting time, money and effort.

The 'phantom' pump occurred because the pump had been removed as part of an upgrading exercise but the plant piping and instrumentation line diagrams (P&IDs) had not been amended to show the change so the pump stayed on the preventive maintenance list.

In addition I discovered that the planners tended to plan the simplest jobs first (because it made it look as if they were making very good progress) leaving the more complex jobs till later – sometimes too late. This behaviour we termed 'reverse planning'.

The above represented a serious shortcoming in the work control system but it still did not account for the fact that the planners could not get the plans and schedule completed on time.

I discovered the real reason when I was reviewing the work list later the same week. The number of jobs on the work list was 328. I checked on the work list that we had challenged earlier in the week – 312 jobs!

I asked the planners the reason for the increase.

One of the planners chuckled, looked knowingly at the other planner, who smiled and then said *'Drip, drip, drip.'*

I asked him what he meant and he replied: *'It's our form of Chinese water torture. We open the work list nine months before the Turnaround and close it four months before but everybody just ignores the closure date. They just keep on requesting work almost on a daily basis. We will get new jobs on the work list every week right up until the start of the event.'*

When I asked the planners why this situation was tolerated they shrugged and replied that that's the way it had always been! Then one of them said: *'It gets worse, there's a budget meeting a month before the Turnaround starts and they always find that the work on the work list is going to cost more than they are prepared to spend so they cancel some of the work – work that has already been planned – it's just a waste of time.'*

I got together with the plant manager, the engineering manager and the two planners and introduced them to the four point gap analysis (see Chapter 15).

It should be noted at this point that the Turnaround manager was not involved in the process as he traditionally 'took up the reins' one month before the event.

We determined the following:

What the client had was a system that almost guaranteed failure via unrestrained job requests which made it impossible to finalize plans, schedule or budget coupled with a reactive cost cutting exercise that disrupted the planning and scheduling that *had* been done a few weeks before the Turnaround start date.

What the client needed was a system that would guarantee that plans, schedules and budget would be completed one month before Turnaround to allow for comment, review and finalization.

Comparison of the two scenarios exposed the following shortcoming with what the client had:

1. There was no budget allocated for the Turnaround before planning commenced. They were in effect 'planning in the dark'.
2. There was no means of challenging work that came onto the work list. They were planning unnecessary and duplicated work.
3. The work list closure date was ignored by everyone. They were trying to plan against a moving target.
4. The 'cost cutting' exercise was performed far too late in the day and disrupted the planning and scheduling and disorganized the start of the Turnaround.
5. The status quo was accepted by everyone as 'the way it is done'. Some people moaned about it but there was no focused attempt to change.
6. Most important, because the management took little interest in the planning of the Turnaround the planners had no authority to enforce a more disciplined approach. This was exacerbated by the fact that, until one month before the start of the Turnaround, there was in effect no one managing the Turnaround preparation phase.

Step 3 – Action to improve

A meeting was arranged with the production manager, engineering manager, Turnaround manager, the two planners and me. They formed an impromptu steering group with an agenda to sort out the work control problems.

I made a presentation to the group that emphasized the fundamental importance of a controlled work list, stressing that the Turnaround was a 'task centred event and every other element of the Turnaround depended upon an effective work list. I introduced them to the computer acronym GIGO (garbage in > garbage out) which seemed to fit the situation nicely.

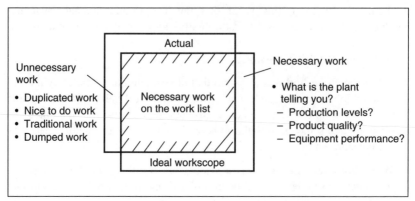

Figure 17.1 Effective workscope

I took them through an 'ideal workscope' exercise detailed in Figure 17.1 using the gap analysis approach to demonstrate what was required.

Then the following action plan was drawn up – see Figure 17.2:

The use of this form instilled discipline into the work order system. Once everyone had been briefed on the importance of work list control and made to understand that they would have a hard time convincing the plant manager that, unless the request was for a genuine defect that had occurred after the work list closure date, there was a good reason for the job not being on the work list.

There was a rush of requests in the two weeks before work list closure but this was reduced to only a handful after the closure date allowing the planners to plan the work rationally and produce a schedule.

- A Turnaround manager to be appointed immediately and take up his post on Monday next.
- A steering group to be set up immediately consisting of plant manager, engineering manager and Turnaround manager, to meet weekly until all Turnaround issues have been resolved.
- Steering group to identify the sum of money to be allocated for the Turnaround to avoid late 'cost cutting' exercise.
- A work list control system to be set up with the following features:
  - Plant manager to inform everyone in writing that the new closure date for the work list is to be two weeks from the date of the letter. The closure to be strictly enforced.
  - Late work authorization system to be initiated from the work list closure date (see Figure 17.3).
  - Weekly work list meetings to be initiated between planners and work requesters to rationalize the work list.
  - Work list challenge routine to be initiated (see Figure 17.4).
  - Necessary work routine to be initiated (see Figure 17.5).

Figure 17.2 Action plan to improve work list control

The person requesting this late addition to the work list must submit it personally to the plant manager		
Order No.	Plant area	

1. What is the nature of the work being requested?

2. What is the estimated cost of the work?

           Manpower   Cost of materials and services

3. Why was this job not on the original work list?

4. What would be the consequences of not doing this job?

I approve/reject this late work request (delete as appropriate)

Plant manager's
signature...Date...

Figure 17.3 Late work authorization form

The work list challenge (Figure 17.4) was performed on every job on the work list. The necessary work routine (Figure 17.5) was a single exercise.

The outcome

The outcome of the above actions was that the work list stabilized allowing the planners to organize and complete the planning and scheduling two weeks before the start of the shutdown. This gave time for a risk assessment to be carried out on the schedule and greatly increased the team's confidence regarding a successful outcome to the event.

The other advantage was that once the work list stabilized the Turnaround manager was able to produce a 'ballpark' estimate (accuracy + or −20 per cent). This indicated that the budget allocation was 25 per cent less than required. The senior management agreed to provide the extra funds because the plant manager was able to prove to their satisfaction that, by using the work list control system, he has justified every job on the work list.

Work order no..Plant location............................. Job description	
PURPOSE	
The purpose of this document is to ensure unnecessary work is eliminated from the work list. Answer the following questions as indicated – for Yes/No questions circle the appropriate response.	
1. What type of benefit will be gained from doing this job?	Hazard reduction Reliability Life extension Improvement Compliance
2. What is the estimated man-hour cost of this task?	
3. What is the estimated material cost of this task?	
4. Is this job a duplicate of some other job on the work list?	YES NO
5. Could this job be eliminated from the work list without compromising safety of production?	YES NO
6. Can this job be done at some other time?	YES NO
6. Is there any other job or project work that could eliminate the need for this job?	YES NO
7. Can this job be simplified?	YES NO
8. Can this job be combined with any other job save time or expense?	YES NO
9. Can a new technique or technology be used to save time or expense on this job?	YES NO
10. Can the need for this job be designed out of the system?	YES NO
11. Is there any other information available that would influence the need or ability to do this job?	YES NO
Comments (if the answer to any of the questions 4 to 11 is Yes)	
This job is to be included on/eliminated from the work list (delete as appropriate). Authorized by: Name...................................Signature...........................Date.................	

Figure 17.4 Worklist challenge routine

PURPOSE
The purpose of this procedure is to ensure all necessary work has been identified.
Review each of the cases below.

---

1. Have RBI programmes been completed and 'minimum required' work lists generated?
2. Has a plant survey been performed to identify any indications that the plant is not performing properly due to some fault that needs to be rectified, including but not limited to the following:

   (a) Product throughput reduced due to constrictions or blockages.
   (b) Product quality compromised by contamination or equipment malfunction.
   (c) Unreliable or malfunctioning equipment.
   (d) Chronic problems requiring root cause analysis.
   (e) Containment problems (indicated by leaks or spillage).
   (f) Control problems requiring investigation/resolution.

3. Is it possible to 'design out' any chronic equipment or production problem?
4. Is there any item of equipment not currently on the work lists that has caused problems on past Turnarounds during plant start-up – if so, should it be included on the work list?
5. Are there any opportunities to eliminate any feature or element of future shutdowns?

Any issues identified by the above review should be appropriately actioned by:

- Submitting a job request for new work
- Further testing
- Referring the issue to engineering department

Figure 17.5 Necessary work

18

Case study 3 – Contract strategy

Step 1 – Presenting problem

Client No. 3 had a Turnaround cycle of four years. Two years after their previous shutdown there had been a significant downsizing that had reduced production and maintenance staff to the minimum required to run the plant on a day-to-day basis. After a fairly difficult year of adjustment the management and staff had settled into a fresh routine and were fairly comfortable with the new situation.

The company was facing a major Turnaround in 12 months' time. The management had had one steering group meeting at which it was determined that a new contractor strategy was required. The old strategy of simply hiring contractors to supplement company staff, who planned and executed the Turnaround, was no longer viable as there was not enough staff.

The company was at a loss as to what their new contractor strategy should be as they had no previous experience at dealing with contract companies at any higher level than hiring 'bodies' to perform 'shop-floor' work.

The company had also shed all of its supervisors in favour of 'self-managed' teams of technicians for maintenance.

In this case, from the point of view of gap analysis, the management knew 'what they had' but they had no clear idea of 'what they needed'. To them the options seemed endless. They had discussed different options but had become bogged down because they had no agreed method of determining the relative costs and benefits of each option.

I was contacted to act as facilitator to assist the management team in the choice of a contractor strategy that would suit their specific needs and culture. Using the principles of the Model of Excellence I designed a routine that would assist the client to define 'What they needed' – it consisted of the following steps:

Step 2 – Investigating the issue

I interviewed the management team and it rapidly became apparent that there were some basic issues constraining their progress.

Using what I have come to call my 'Provocation Technique' I asked them a simple question:

'Why is it difficult for you to change?'

This always provokes a specific response beginning with (or implying) the word 'because' and often leads to the heart of the matter.

I outlined the team's main responses in writing in the form of issues and underlined the key words which I took to be their criteria for judging whether a particular organizational combination suited them. The responses to the question 'Why is it difficult for you to change?' were:

1. Because they had had a bad experience with a contractor two years before on a large capital project (their statement was 'we got burned') and they were still smarting from the experience. The contract had gone badly and generated two negative features. The first was that the company felt that they had lost <u>control</u>, and the second was that they did not trust contractors to work in their best interest.
2. Because they had had only bad experiences from their dealing with contractors they were worried about the <u>contractual difficulties</u> involved in drawing up an unfamiliar type of contract (for a Turnaround).
3. Because they had had doubts about their ability to control the <u>cost</u> of the event if contractors were involved.
4. Because they felt that using contractors on a Turnaround would make it a more <u>complex</u> event than if it had been in-house.
5. Because they were concerned that dealing with an external organization would lead to <u>communications difficulties</u>.
6. Because they didn't know how <u>difficult</u> it would be in general to change from what they had done in the past to what they needed to do in the future.

Step 3 – Generating criteria

The above deliberations gave us a foundation to work with because, although they felt that they had no criteria, the exercise had actually generated the criteria (underlined) that were important to this company (different companies generate different criteria, depending on their situation and culture). The company's specific criteria were:

- Cost
- Loss of control
- Communication difficulties
- Complexity for client
- Contractual difficulties
- Difficulty for client (general)

These criteria represent what is most important to the client when considering the issue of formulating a contractor strategy that would suit their specific needs and comply with their cultural norms.

Step 4 – Generating options

In order to differentiate between the various contractor approaches we required a relatively straightforward way of defining the critical features of the client/contractor relationship. After some discussion we settled on generating the options by asking three questions:

- Who will manage the Turnaround?
- Who will supervise the work?
- Who will provide the human resources?

Step 5 – Generating the matrix

The different options and the client's criteria were set in a matrix so that the client's attitude to each option could be measured against each of the criteria using a simple 1–10 scale of scoring. In this case, because of the way the criteria were stated, the lowest scoring option would be the favoured option.

Figure 18.1 shows what the matrix looked like.

				Score each box between 1 and 10: 1 = Low, 10 = High			
Criteria Contract strategy	Cost	Loss of control	Communication difficulties	Complexity for client	Contractual difficulties	Difficulty for client	Score

1. Client manage
   Client supervise
   Client resource							

2. Client manage
   Client supervise
   Contractor resource							

3. Client manage
   Contractor supervise
   Contractor resource							

4. Managing agent
   Contractor supervise
   Contractor resource							

5. Contractor manage
   Contractor supervise
   Contractor resource							

6. Integrated team
   (Select 'best' from
   each organization)							

7. Any other?							

Figure 18.1 Example of contractor strategy matrix

Option 1

This would have involved the client doing the Turnaround with only internal management, supervision and resources drawn from their own plant and other plants in the company.

Option 2

The client provides management and supervisors to plan, prepare and control the execution of the Turnaround using resources that are drawn from agencies (known in the trade as 'body shops').

Option 3

The client appoints a manager only and all supervision is provided by one or a number of contractors. The manager is in charge of the day-to-day planning, preparation and execution of the Turnaround.

Option 4

The client hires a professional Turnaround/project manager (instead of appointing one of their own managers) to act on their behalf as an agent who will manage the contractor supervision and resources and is in charge of the day-to-day planning, preparation and execution. The managing agent would form the link between the client and the contractor.

Option 5

This option is sometimes called a turnkey contract, where theoretically the client's only involvement is to specify the workscope, negotiate the contract, monitor the contractor's performance and pay the money on completion of the work. Often, when this option is used, the client provides a 'project manager' who is the single point liaison between the client and contractor for the Turnaround but does not get involved in the day-to-day work.

Option 6

This option involves the client sitting down with the contractor (or a number of contractors) well in advance of the Turnaround (9–12 months) and jointly building an integrated team from client and contractor personnel using the principle of selecting the best person available for each key role (this assumes of course that the organization has been designed).The idea is to put together the strongest possible team and then give them enough time to form, storm and norm so that they can perform effectively. This does not normally include tradesmen unless they have a critical role to play.

Step 5 – Scoring criteria against options

Each member of the team was asked to fill in the matrix individually, scoring each criterion for each option with a score of between 1 and 10. The lower the score is for each criterion, the higher the confidence level.

Criteria / Contract strategy	Cost	Loss of control	Communication difficulties	Complexity for client	Contractual difficulties	Difficulty for client	Score
				Score each box between 1 and 10: 1 = Low, 10 = High			

1. Client manage
   Client supervise
   Client resource
	—		Not feasible			—	—

2. Client manage
   Client supervise
   Contractor resource
	3	3	7	8	5	8	33

			Difficult to achieve				

3. Client manage
   Contractor supervise
   Contractor resource
	4	4	5	6	5	6	30

4. Managing agent
   Contractor supervise
   Contractor resource
	6	6	4	5	5	6	32

5. Contractor manage
   Contractor supervise
   Contractor resource
	5	8	4	3	4	3	27

6. Integrated team
   (select 'best' from
   each organization)
	3	5	3	3	3	2	19

7. Alliance	—		Not feasible			—	—

Figure 18.2 The matrix filled in

It is important to note here that these scores are not finite scores (i.e. exact values based on some predetermined scale) but comparative scores that indicate an individual's preference for one option over another.

Once the individuals had completed their matrices the scores were aggregated in a master matrix (Figure 18.2). Any large discrepancies between an individual score and an aggregate score were discussed and resolved by the group to ensure a common approach to the strategy. Although there were some individual discrepancies, all members of the team opted for option 6, for the following reasons:

Option 1
This was not feasible as the company did not have enough personnel to operate it.

Option 2
This could be a feasible option but it would put too much strain on the client's organization and would mean that while people were involved in the Turnaround, many other company functions would suffer. It was considered to be too difficult to achieve.

Option 3
This would be difficult for the client as there was no one in the client's organization who had the specific skills required to manage the Turnaround.

Option 4

This option would have overcome the difficulty detailed in option 3 but was not favoured because it would have involved an additional agency and the cost of the managing agent would have been high.

Option 5

This option was culturally unacceptable. The team could not support giving full control to the contractor (this is where the question of trust came to the fore).

Option 6

This was the chosen option for the reason that it would address the two issues of insufficient client personnel and desire to retain control. Time would be needed to define the relationships in the integrated organization but it was felt that 12 months was enough time to do this.

Option 7

Reality is open ended so the team considered other options such as alliances and partnerships between themselves and the contractor. It was felt that there was not sufficient time to create either of these contract strategies in time for the next Turnaround.

The outcome

The client went forward with the 'integrated team' option and:

- Over the 12 month planning and preparation period developed a very progressive relationship with the two contract companies it had selected to fill the appropriate roles in the organization.
- In order to ensure that everyone understood what was required they developed a set of roles and responsibilities jointly with the contractors.
- During the period there were a number of 'get-togethers' to resolve issues and suggest improvements in approach. The final act that cemented them together was when they settled on 'One Team' as their slogan.

After the Turnaround was completed the client stated that generally they were very satisfied with the way the organization had performed. Although they had delegated many of the key roles to the contractors, the client felt that, at all times, they were in control of the Turnaround. Because there had been buy-in from all parties and a common objective, the client had come to trust the contractors much more than they had in the past. They considered the One Team concept to be a great success.

The two contractors also had a positive attitude to the experience. Obviously they would have preferred to have been given the complete contract but they both stated that they had learned many useful lessons about client/contractor relations. They also judged the One Team concept to be a great success.

19

Case study 4 – Cost control

Step 1 – The presenting problem

Client No. 4 had undergone a change of management and the new business manager was very cost conscious. He had required the managers of each department in the company to submit an outline budget for the next two years. This was so that he could get an idea of the spending patterns within the company as a first step to controlling and then reducing expenditure.

The maintenance manager submitted his budget for day-to-day maintenance but there was nothing included for Turnarounds. He explained that the money for Turnarounds normally came from another fund. He was then asked by the new business manager to give at least an account of the expenditure on Turnarounds. The maintenance manager replied that he could not do so because no detailed accounting was ever performed on Turnarounds. The company's traditional approach had been '*the Turnaround will cost whatever it costs*' so that whatever money was needed was provided.

This attitude did not sit well with the business manager and he instructed the maintenance manager to change his system. The business manager wanted three things

1. An estimate of costs for the next Turnaround (nine months hence).
2. A cost accounting system to record expenditure before and during the Turnaround.
3. A strategy for reducing Turnaround costs (he reckoned that if the company had never measured its costs then it was odds on that they were overspending).

The maintenance manager asked that an external consultant be engaged to facilitate this change because neither he nor any of his staff had any kind of experience of cost accounting a Turnaround. I was engaged to facilitate the change.

Step 2 – Investigating the issue

At the first steering group meeting with the business manager, production manager, maintenance manager and the appointed Turnaround manager, it was made pretty clear that, in terms of gap analysis, the business manager knew **'what he had'** (a total lack of cost control) and **'what he wanted'** (a cost control system for Turnaround costs), but I was not sure if he knew **'what he needed'**.

I asked the team (as I always do) if their interest in controlling costs was confined to the direct costs of the Turnaround or if they had considered the 'total cost' of the Turnaround as part of a wider strategic consideration.

None of the managers understood the concept and the business manager asked me to explain. I did so, in the following manner.

Direct costs

To find the direct cost of a Turnaround you add up the all of the money expended for manpower, materials, equipment and services during the preparation, execution and termination phases of the Turnaround.

Total costs

To find out the total cost of the Turnaround it is useful to ask the question:

'If we did not perform the Turnaround, how much would we save?'

This involves much more than the direct costs. Consider Figure 19.1.

The total costs are made up of:

The direct costs

These we have already dealt with. What can be added here is that the direct costs are considered to be part of production costs.

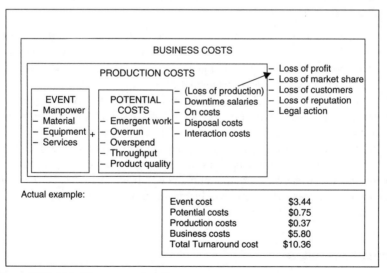

Figure 19.1 Total Turnaround costs

Potential costs

Also contained within production costs, these are all of the unplanned costs that can (and often do) occur. Emergent work is, as its name implies, unplanned work that only emerges after the start of the event. Over-run means taking longer than planned with the consequential administration costs of running the event for longer.

Overspend is simply spending more money on the planned work than was intended. Throughput refers to the potential interruption to or restriction of the product throughput due to an ineffective plant start-up. Product quality can also be affected by an ineffective start-up.

Production costs

The most significant cost in this element is profit lost through the plant being shut down. This cost often dwarfs the direct costs of the Turnaround but surprisingly is often not considered by companies to be part of the Turnaround costs. The consequence of this is that the two costs are kept apart. This can lead to an illogical situation.

> **Example**: If the emphasis is only on keeping the direct costs of the Turn-around to a minimum and total costs are not considered, a proposal to spend an extra £35,000 to carry out a major piece of work in a more effec-tive way that will cut half a day off the duration of the event will most likely be rejected because it significantly increases the direct cost of the event. However, when we consider that the plant makes £210,000 per day profit it seems illogical to reject a proposal that will offer an immediate 3 to 1 return on investment!

Downtime salaries of production personnel who will not be gainfully employed during the Turnaround should be considered a Turnaround cost. On costs are the costs borne by the company if it needs to buy product on the spot market to re-sell to a customer in order to retain that customer.

Disposal costs, especially for companies that produce toxic effluent, are increasing each year as health and safety rules prohibit simple dumping of effluent.

Interaction costs are those that are generated by the interaction with plants upstream (which normally supply raw materials or feedstock) and downstream (which normally receive the output of the plant). If the plant is to be shut down for say three weeks, what costs will the upstream and downstream plants have to bear because of this?

Business costs

Business costs can be regarded as the accumulation of all of the included costs. For example, loss of production translates into loss of profit at the business level.

The other business costs are more ephemeral but their consequences can be substantial. Delays in starting up the plant lead to unplanned loss of production and an inability to service customer's needs. This could result in a loss of customers and/or a loss of market share. This area is so sensitive that many large companies are obliged to inform the stock market when they are planning major Turnarounds.

And finally, the thing companies dread above almost anything else is loss of reputation due to some error, especially if that error results in litigation and public censure. These costs are difficult to evaluate except when they occur and then they are all too apparent.

Figure 19.1 also gives the actual difference between direct costs and total costs for this case. In this example a factor of 3 to 1, but the factor can sometimes be much larger.

Once business managers understand the financial impact a Turnaround can have on profits they tend to take Turnaround cost control a good deal more seriously.

Step 3 – Formulating the cost strategy

In this case, the steering group agreed that, from a business standpoint it would be to their advantage to employ the total cost model. The decision was taken to set up a single Turnaround cost centre and code number and all costs that were recognized to be Turnaround costs were booked to that number. This included management time – including the time spent at steering group meetings and:

- Every hour that was spent by anyone on the planning, preparation, execution or termination of the Turnaround;
- Every item of materials used on the Turnaround;
- Every item of equipment used on the Turnaround;
- All support services;
- All infrastructure costs;
- Any incidental costs that could be directly attributed to the Turnaround;
- A penalty sum to cover the loss of production/profit.

Having set the context for cost control, the steering group then authorized the Turnaround manager to engage one of the company's cost accountant/quantity surveyors to act as 'cost co-ordinator'. This person would work with the Turnaround manager and myself to tailor make a cost control model to suit the specific needs of the Turnaround.

Step 4 – The rational cost model

As we were starting from scratch on the cost control model we were able to construct a rational model that would identify all the different categories of direct costs associated with a Turnaround.

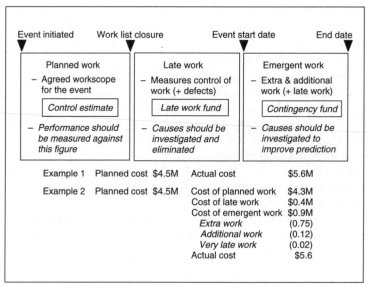

Figure 19.2 Rational cost model

A brainstorm session was arranged which developed what came to be known as the 'rational cost model'. Figure 19.2 shows the final output of the session.

The idea behind this model came to be known as 'cost mapping' because we were literally creating a template for a detailed map of the overall Turnaround that would identify (in planned and actual spend terms) how every penny was spent.

We started by identifying the four critical cost dates on the Turnaround timeline. They were: the date that the event was initiated (first costs – the first steering group meeting); the work list closure date; the event start date and the event end date.

Control estimate

The period between the initiation date and the work list closure date contained the agreed workscope for the Turnaround and would be used to calculate the control estimate for the event. This was the figure against which the Turnaround team's cost control performance should be measured.

Late work fund

All of the work that was requested after the work list closure date up until the start of the Turnaround (apart from defects that had been discovered since the closure date) would be accounted for in a separate fund because these jobs were actually a measure of the company's inability to control the work list. The root causes for this inability would need to be found and action taken to eliminate this shortcoming. (The use of a late work authorization system as detailed in Chapter 17 would go a long way to defining this situation.)

Contingency fund (for emergent work)

During the period of the shutdown itself work 'emerges' that was not planned for and, in my experience, will typically add 5–15 per cent to the planned cost of the event. This work needs to be identified, categorized and costed in a separate fund. The categories are:

Extra work Work generated by an existing work order (an inspection exposes an unexpected defect that needs to be rectified).

Additional work Work generated by a new work order during the event (e.g. a pump fails on start-up and needs to be replaced).

Very late work Work that should have been on the initial work list but was ignored or forgotten till the event started.

Very late work is treated the same as late work. Extra work is investigated to ascertain if it could have been predicted and if it is likely to occur on the next Turnaround, in which case it can be added to the work list. Additional work is recorded and put onto the work list for the next Turnaround so that it can be considered for inclusion on the next work list.

It is only by measuring every item of expenditure, including the unplanned work, that it is possible to exert real control over Turnaround costs. Figure 19.2 also shows two examples of reporting costs.

Both examples are reporting on the same event. In the first, all that is known is that there was a £1.1M overspend. With no measurement available there can be no analysis or improvement. In example 2, using the rational cost approach, the overspend is broken down into different categories of spend (backed up of course by detailed figures) which can be analysed and action taken to improve performance).

Step 5 – The cost breakdown structure

We now needed to cast the principles of the rational cost model into a more detailed model which we named the 'Cost Breakdown Structure'. Figure 19.3 shows an example of the model we developed (the actual model was more detailed).

We broke the plant down into three geographical areas, each of which would have its own budget and estimate. We further broke each area down into the categories shown on the left-hand side of the diagram. They have

Figure 19.3 Example of a cost breakdown

already been defined apart from 'Infrastructure' which consisted of all of the temporary cabins, cabling, plumbing, phone cables etc. that needed to be installed for the Turnaround.

Costs for each category were estimated and reported against individual work orders so that at the end of the event we would have an accurate cost for all different types of work. The costs were also accumulated by type so that we would have a picture of the overall patterns of expenditure.

This approach gave an almost infinite array of estimating and reporting options from broad brush stroke to very detailed costing. It also allowed for comparison of similar jobs in different areas of the plant to ascertain why one cost more than another. The fact that we had a 'planned' cost map and an 'actual' cost map meant that we could identify discrepancies between them (one example was – in Area A we had a planned expenditure of £43K for scaffolding whereas the actual spend was £57K).

Using the two maps we were able to identify where we had overspent, where we had underspent, and where there were patterns of common faults such as corrosion that had generated the extra cost of repair.

Step 5 – The cost model

The final step in the process was carried out by the cost engineer who constructed a specific cost model for the Turnaround in SAP. A good deal of work was required to develop the model but once it was finalized it could be used for all Turnarounds in the future.

Outcome

The model served the company well and allowed them to significantly alter and improve their approach to Turnarounds and brought all expenditure under the control of the management.

20
Case study 5 – Logistics

Step 1 – The presenting problem

Client No. 5 had a history of over-running the durations of Turnarounds, in some cases by a number of days (losing production and profit). Their gut feeling was that the over-runs had less to do with ineffective planning or execution of the actual work and more to do with the fact that vital components seemed to either go missing or arrive too late to be installed on the planned dates. But it was just a feeling because they had no means of measuring the situation.

This is not an uncommon situation. Often excellent technical planning is let down by sloppy logistics planning. One reason I have found is that some companies are unaware of the importance of logistics planning.

What the client seemed to be describing was a logistics problem involving bulkwork which, as defined in Part 1, is comprised of the many small items of equipment (valves, pumps, motors etc.) that need to be removed from the plant and either replaced by a new or overhauled spare or sent for rectification and then returned to the plant so that they are available for re-installation when they are required. On a major event there may be many hundreds of these items and every one must be properly organized.

Step 2 – The assessment

When performing an assessment on a situation where there is no formal system and nothing is written down, the only information available is anecdotal – it is all you have to work with.

I have found that the best way to tap into people's memories of how an informal system works is first to ask *positive* questions, e.g. 'What are the steps necessary to achieve this?' or 'How does the process work?'

The answers to these questions will tell you how the system *should* work.

Follow the positive questions with *negative* questions, e.g. 'What are the things that really annoy you about this situation?' or 'What gives you the most trouble?'

The answers to these questions will tell you how the system *does* work and why and how it fails.

because they tap into the real pain, anger, and frustration that people feel when the system prevents them from doing a good job.

During interviews with company personnel it became apparent from the anecdotal evidence that the problem was one of 'disorganization' which is the term I use to describe a situation where a lot of people are working hard and doing their best but, because there is no centralized co-ordination, communication or responsibility, things tend to proceed in a somewhat haphazard manner.

A disorganized system can often deliver 90 per cent of what is required of it (albeit at a higher cost and with greater effort) but it is the 10 per cent it cannot deliver that causes the problems. If ten components need to be overhauled offsite and only nine of them are returned to site on time, the missing component suddenly becomes a constraint on duration because it will delay the plant start-up.

Although a single valve in a Turnaround may be considered to be a small and insignificant job, if it is not available when it is required it can enforce a financial impact way beyond its direct cost (for example, the direct cost of overhauling the component is £1500 but the consequential loss of profit one day's lost production caused by its delay is £150,000 plus a potential loss of business and reputation).

In this case the client's personnel were well versed in the formal planning and execution of the technical work in the Turnaround but lacked the ability to 'close the loop' and ensure that everyone did what they said they would from a logistics point of view.

There was no central control of logistics – it was divided between eight different people, who were only concerned with their own part of the job.

A second shortcoming revealed was the lack of a plan for bulkwork logistics. As mentioned before, a single component taken in isolation may be considered a pretty simple job but when you have many components and each has to be treated in a particular way then you have a complex situation. The client was ignoring this complexity (although he had many hundreds of different bulkwork items to deal with) and this was seriously diminishing the ability to control the event.

The assessment discovered that the issue of logistics had never been discussed in any integrated fashion.

Step 3 – Defining the problem

Our gap analysis exercise showed that the client:

- Could not describe the current situation in any coherent way so did not know **'what he had'** (except to remark that it was 'sometimes pretty chaotic');
- Knew **'what he wanted'** only in the most general terms (i.e. all bulkwork items to be available when required);
- Had no idea of **'what he needed'** to close the gap.

Clearly, some method was required that could first define the complexity of the situation and then seek to control it.

Step 4 – Defining complexity

A comprehensive analysis and action programme was instituted to deal with the logistics complexity. What follows is one portion of the programme but it demonstrates the thinking behind logistics planning.

The client was in the chemical industry and so what was considered to be one of the most mundane jobs of a Turnaround – the isolation valve – will be used to describe the process. It should be noted that this was only one of a number of programmes going on at the same time, some of the others being: control valves, safety relief valves, small pumps, small electric motors etc.

First, using information from the Turnaround work list, the preventive maintenance routines and defect sheets, we defined the requirements and the complexity of the isolation portion of the programme as follows:

- There were 67 isolation valves to be dealt with;
- 53 of the valves needed to be decontaminated;
- 9 of the valves were to be replaced with new valves and scrapped;
- 33 of the valves were to be overhauled and re-installed;
- 25 of the valves were to be replaced by previously overhauled spares and then to be overhauled after the Turnaround and returned to the stores as spares;
- A local offsite workshop would be engaged to overhaul 29 of the 33 valves during the Turnaround and 25 valves after the event;
- The remaining 4 valves were 'specials' and would be sent to another workshop 150 miles distant from the plant.

Normally, each supervisor was responsible for his own valves (as well as all of his other duties). This resulted in multiple interfaces with the workshops where the valves were overhauled and with the stores from which new and spare valves were issued. It also generated a competitive atmosphere in which each supervisor considered his own valves to be top priority and made efforts to ensure his valves were overhauled first, sometimes to the detriment of other areas.

An interview with the workshop manager revealed that during the course of a Turnaround there would be many emergency phone calls and arguments about which valves he should be working on at any particular time – leading to his priorities being regularly changed. This aspect of the disorganization cost the client money and sometimes generated delays in the overhauling of genuine priority valves.

There was a roughly defined area on the plant where valves were dumped on their way from the plant to the workshops and on their way back. This led to some confusion regarding 'who owned what' and where things were. More than once this had resulted in a valve that had been removed from the plant a few days before being re-installed before it had been overhauled. There had also been a number of occasions when contaminated valves had found their way to the workshops, only to be sent back as being in an 'unfit condition'. This latter issue had generated safety concerns and on one occasion resulted in the involvement of an HSE inspector. On one Turnaround, two valves had simply disappeared and were never found again.

Step 5 – Action plan

After careful analysis of all of the elements in the bulkwork scenario the following was agreed as the outline for an action plan with the following elements:

Appointment of bulkwork marshal

It was agreed that one of the supervisors would be appointed as bulkwork marshal who would have sole responsibility for the movement and control of bulkwork. This reduced the number of supervisors available to supervise work but it was estimated that the number of hours saved by each supervisor not having to be involved in logistics would more than compensate for the shortage.

Use of control sheets

- The supervisors would be issued with schedule sheets for the valves in their areas (see Figure 6.5 in Part 1) to define the planned requirements for each valve;
- A control sheet (see Figure 6.6 in Part 1) would be displayed in the control cabin showing the required logistics stages for each valve (to be filled in by the relevant person responsible for each stage);
- A valve marshalling (see Figure 6.7 in Part 1) which would stipulate on which day each of the valves would be removed from the plant by the work teams and be made available to the valve marshal and then specify the latest acceptable date on which the valve (or its new or overhauled spare) must be made available to the supervisors for re-installation on the plant.

The logistics plot plan

What was needed to demonstrate how the complexity could be managed was a diagram that would identify the route for each of the different categories of isolation valve (as well as all other bulkwork). A team consisting of the Turnaround

Figure 20.1 Marshalling bulkwork (isolation valves)

manager, the newly appointed bulkwork marshal and myself performed a brainstorming exercise. The output of the exercise is shown in Figure 20.1.

It was only after we had laid out this diagram that the client's people started to realize how complex the logistics of bulkwork was and believed they had found the reason why it had so often gone wrong in the past.

Consider the complexity involved when all of the other items in the logistics programme (control valves, relief valves, pumps, electric motors etc.) are added to this, the number of workshops involved is increased to seven (one onsite and six offsite) and the additional complication when the system is operated by eight individuals who are only concerned for their own items.

Complex processes require careful planning. The following actions were developed:

- Specific areas of the plant were nominated for each for the decontamination and lay down bays.
- The diagram (Figure 20.1) was overlaid on the plant footprint to create a Turnaround plot plan that would show everyone where all of the structural elements were located (item bays etc.) and the routes between them.
- The diagram was also used as a flow diagram that specified the order and sequence of the steps required to move the different categories of isolation valves through the process.

- The valve marshalling sheet (Figure 6.7) would specify the required timings on each individual valve.
- The valve marshal would be responsible for:
  - Producing the logistics plan that tied all of the elements together (his sole responsibility).
  - Liaising with the supervisors before the event to ensure that everyone understood the importance of making the items available on the stipulated dates.
  - Expediting the removal and return of items during the event.
  - Controlling the movement of valves offsite and onsite during the event.
  - Organizing the flow of items through the water wash decontamination bay and checking on the progress of the work on a daily basis.
  - Organizing the timetables and different modes of transportation needed to move the items. This would include marking out a 'one way system' to simplify traffic patterns and the nomination of 'pedestrian only' routes where vehicles would be prohibited to reduce the risk of accidents.
  - Agreeing an overhaul schedule of item delivery and pick-up with the workshop managers to ensure all overhauled valves were returned on due dates and liaising with them on a twice daily basis during the event to report and resolve issues.
  - Regularly reporting progress and problems to the Turnaround manager and especially any issues that could only be resolved by the Turnaround manager.

The supervisors, who would be responsible for providing the input into the bulkwork marshal's process and would be the recipients of its output, were extensively briefed on their part in the process. Everyone bought into the process because they could see its value.

Outcome

The bulkwork marshalling programme was implemented on the next Turnaround and proved to be a success. There were no delays due to missing or late items. After the Turnaround the new logistics system was identified as one of the major improvements to the Turnaround approach.

There was a noticeable reduction in 'fire fighting' during the event and the supervisors all agreed that the new system was a vast improvement in their working conditions.

The workshop managers were enthusiastic about the new system as they now only had one person to deal with (one marshal instead of eight supervisors).

21

Case study 6 – SASOL experiences in cost/risk optimization

Introduction

Companies that lead the field in Turnaround management are constantly seeking new solutions to the strategic issues that face them when they attempt to set a workable strategy for Turnarounds, mainly because, for progressive companies, the old solutions no longer serve the need nor satisfy their desire to execute these events safely, within the planned duration and budget and to the required standards. This is the goal of Turnaround management.

Any approach, technique or technology that brings us closer to this goal should be considered on its merits (which are ultimately defined by what it delivers and how effective the client views it) and if it serves the need then it is only common sense that it should be adopted providing it is cost effective.

The fusion of cutting edge software developments with engineering and local plant knowledge can create a formidable enabling tool and provide a sea change in the strategic approach to Turnarounds. The following case study recounts the experiences and outcomes of applying such a new technology, developed by the Woodhouse Partnership, to the issues facing SASOL, a large South African company.

This case study describes some of the methodologies, usage and results obtained during the last four years at the Secunda plant of SASOL. In particular it describes the optimization of the major shutdown strategy, and some of the many individual decisions about what capital investment, maintenance or inspection is worthwhile and when.

SASOL in South Africa is responsible for around 50 per cent of the total domestic fuel and lubricants market, largely through its world-leading coal-to-gas-to-oil conversion technologies. The Secunda site comprises the largest such production facility, some $10 \text{ km} \times 4 \text{ km}$ of integrated process plant, with over 23,000 pressure vessels and a feedstock obtained from five coal mines operating 24/7. In the past, the company benefited from protected political and commercial status but these protections are rapidly disappearing and world markets are providing significant price competition. Accordingly, great pressure has been brought to bear on the unit production costs, product quality and operational flexibility of the plant, and significant investment has been made in new technologies and ways of working.

The Woodhouse Partnership Ltd (TWP) comprises about 40 international experts in operational reliability and asset management (training, facilitation, tools), and works in the energy sector, public utilities, transport, mining and manufacturing industries in around 25 countries. TWP has been working with SASOL for around five years to establish better practices and implement modern asset management processes and tools.

Where it all began

In the 1980–1990s, like many companies, SASOL had tried most of the 'three letter acronyms' (RCM, RBI, TPM, TQM etc.) – and achieved varied, usually temporary successes with some of them. The move towards a more integrated operational reliability approach coincided with the appointment of Bram de Klerk as managing director of the Secunda operation, the researches of Ray Owen (head of shutdown planning) and the first TWP training courses in the subject. It was quickly apparent that this 'integrated roadmap' approach would be far reaching in education, organization structure and in potential benefits. The first pilot studies of component activities (maintenance tasks, spares requirements, problem-solving/root cause analysis) revealed scope for multi-million dollar potential savings or performance improvements. A reliability group was established to co-ordinate and facilitate the adoption of such methods, and a more systematic (criticality targeted) roll-out of RCM and RBI studies ensued. At the same time, a number of further studies were performed using the APT cost/risk optimization tools to explore not just what was worth doing, but the optimal amount, interval or timing of the tasks. These showed further spectacular benefits in cases ranging from turbocompressor inspections and maintenance, valve reliability and design, gas turbine rotor spares, pipeline corrosion monitoring and electrical protection system testing.

The big decisions

In 1999, with the emerging evidence of (a) the practicality (particularly when faced with poor or missing data) and (b) the value of the techniques, some of the critical strategic decisions of plant operations were tackled. In particular, the Synthol business unit, containing the catalytic heart of the gas-to-fuel conversion process, was under pressure to operate at higher uptime, for longer run-lengths. The current two-yearly major Turnaround was being challenged for change to a four-year interval. However, a considerable number of operational, metallurgical and inspection/maintenance issues would need to be considered to enable such a significant step. The business unit manager volunteered his team as a proving ground for a radical new way of evaluating and optimizing the total reliability, performance, cost and shutdown strategy. The following describes this process and the results obtained.

Optimization of shutdown work requirements and intervals

The study considered the top maintenance and risk areas, each modelled using the APT suite of optimization tools to allow exploration of data uncertainty. This modelling identified the optimum maintenance or inspection interval, and the cost/risk penalty for deviating from the optimum. In cases where a four-yearly Turnaround cycle was found to be unachievable (too costly and/or risky), alternative designs, configurations or operating strategies were also evaluated. Such de-bottlenecking revealed considerable scope for cost/risk/performance savings, worth a total of **R10.7 million/year** and representing a payback for the necessary capital investments involved averaging just **three months**.

The cost, performance and risk implications of current two-yearly and proposed four-yearly shutdown cycles were then quantified, and key decision drivers identified. This study revealed that an extension from the current two-yearly shutdown to four-yearly would hold *net* cost/risk improvements worth **R40 million/year**. Combined with the improvements available from the 'debottlenecking' actions identified in individual equipment studies, this represents a total benefit from the study of **R50.7 million/year**. This is based on conservative estimates for total cost, availability and risk effects, and includes 'worst case' projections for the associated risks (so the 'most likely' benefits are even greater).

Component studies performed

The studies considered the key maintenance, performance and risk items in the Synthol trains. Criticality analysis considered the 'risks if shutdowns were deferred' (i.e. the degradation rates) and the task costs (including downtime duration required). The identified critical issues and components included:

- SAS reactors: internals and structural integrity, vessel nozzles
- Hot quench tower: internals, shell integrity, safety relief valves
- HQT pumparound system: valves, pump mechanical seals, pumparound cooler
- Trim cooler: cleaning safety, design, bypass options
- Separator drum relief valves
- Overall train performance and reliability
- Shutdown total work co-ordination, duration, impact.

Other items and systems were also considered and examined for their possible effect on the shutdown strategy, but were rejected for a number of reasons (low cost/criticality, maintenance can be done on-line, installed redundancy, or deterioration timescales already permit four-yearly or longer maintenance intervals). Such items included the process pipework, instrumentation, pumps and motors, feedgas compressor and condenser, reactor pressure release valves and fin-fan coolers.

Analysis methodology

The study was performed over a four-week period by two persons full-time and four persons part-time, conducting a series of structured interviews and small multi-disciplined team reviews of component problem areas. In each case, a two-stage modelling procedure was employed, based on the European MACRO project methodology (see Figure 21.1).[1] This provided a structured format for range-estimating the costs, degradation patterns and effectiveness of alternative maintenance, design and operating scenarios. APT-MAINTENANCE, APT-INSPECTION and APT-PROJECT software tools (see Figures 21.2, 21.3 and 21.4) (whose development was commissioned by the MACRO project sponsors) were used to calculate instantaneously the total business impact of each option, allowing immediate 'what if?' exploration and sensitivity analysis (identifying which assumptions

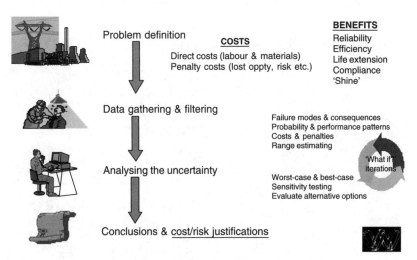

Figure 21.1 MACRO cost/risk optimization process (© The Woodhouse Partnership Ltd 2001)

---

[1] European MACRO project: best practice asset management decision-making, see www.twpl.co.uk

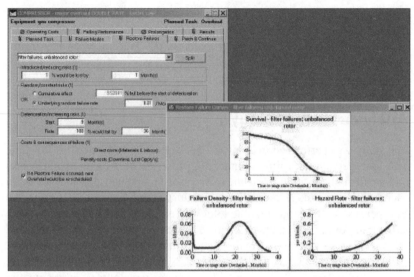

Figure 21.2 Typical reliability pattern estimation (APT-MAINTENANCE)

were critical and worth further research, and which made no difference to the optimal strategy) – see Appendix at end of chapter.

The first problem definition step is clearly very important. MACRO guidance in this area is a valuable way of avoiding many of the practical

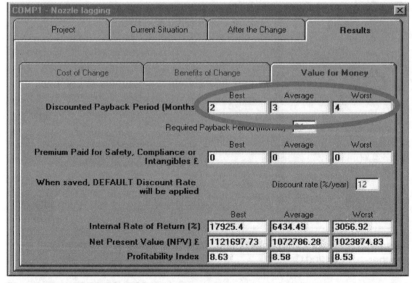

Figure 21.3 APT-PROJECT evaluation of investment (nozzle lagging), including sensitivity analysis to uncertain assumptions

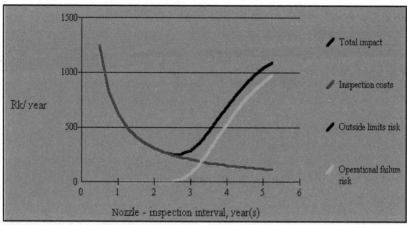

Figure 21.4 APT-INSPECTION: cost/risk trade-off for nozzle inspection interval

pitfalls: for example, in the estimation of degradation rates when hard data is incomplete or of inadequate quality. In such a case, the questions focus on range estimating the *cumulative effects* of degradation/failure (i.e. the survival curve) rather than trying to quantify the 'rate of change of risk' (hazard rate curve) that is really needed for a commercial evaluation. APT software readily converted these survival projections into the implied probability patterns, and allowed us to explore different extremes of opinion in real time (as was obliged in step 3 anyway – to find which assumptions are critical to the decision and where uncertainty is worth investigating further).

The final step is also important – converting the analysis and technical descriptions into a common language: business impact in $$ values. Only then will a proposal be comparable with other, widely different options (expressed in the same value-for-money format). Moreover, financial managers will understand the costs, risks and benefits in a language they recognize!

In the case of activity *timing*, the commercial justification involves quantifying (and identifying the optimal combination of) capital and operating costs, operational efficiency, reliability/risks and planned/unplanned downtime. This involves very complex maths (including renewal theory, reliability engineering and modelling the *interaction* between individual failure modes, efficiency levels etc.) – again the software enabled real-time modelling and 'what if?' exploration.

Example component study: pilot operated relief valves (015)

The HQT safety relief valves are in four critical service locations. Operating at high temperatures, there have been a number of failures of the pilot unit, and seal degradation has been discovered. If the pilot unit fails, the main

Figure 21.5 Total impact curves (best, likely and worst interpretations) for 015 PORVs

valve seals are often damaged, and leaks to flare may result. The valves are maintained by valve workshop, full overhaul every two years, and our analysis considered the risks associated with extension to four-yearly maintenance. When it became clear that resulting unreliability would be very expensive (see Figure 21.5), alternative design options were considered. The possible design improvements included

- Change of seals from Viton to Kalrez material (very expensive but can tolerate the product attack) (see Figure 21.6).
- Installation of a second pilot unit on each valve, allowing on-line swapover and maintenance in event of pilot failure (see Figure 21.7).
- Inspection plan.

It was thought that a robust risk analysis would be needed to obtain permission to extend the maintenance interval for the 015 valves (currently *required* to be maintained every two years). The identical, but less critical, 024 valves on the separator unit may already be allowed to extend to six-yearly. The base study (current design) showed the following cost/risk profile (sensitivity testing, showing worst case/best case/most likely interpretations).

Installation of Kalrez seals

The cost of the Kalrez material is extremely high (R93,000 per set, compared to current Viton costs of R4800). Nevertheless, the longer life/higher reliability provides a payback for its use if maintenance intervals are extended. At the current two-year interval, the high cost/low risk option of Kalrez is about the same total impact as the low cost/higher risk option of continuing with Viton.

Figure 21.6 Viton versus Kalrez seals (PORVs on hot quench tower in 8 m SAS train)

Installation of second pilot unit on each valve

The installation of a second pilot unit, allowing on-line changeover, shows an excellent payback for the one-off costs involved. The total capital costs are estimated at R16,000 per valve, benefits of *c*.R12,000/valve/month, and the payback period for this investment is less than three months. With two such

Figure 21.7 Current versus better seals versus dual pilot options

valves per train, and a total site population of 24 units, the net total cost/risk improvement is worth **R3.4 million per year** if the Turnaround cycle is extended to four-yearly.

Summary results of the key studies

The other studies were performed in a similar manner, with periodic tasks evaluated/optimized using APT-MAINTENANCE (pm work) and APT-INSPECTION (inspections and testing work). Other capital projects to address the shutdown cycle 'bottlenecks' were evaluated using the APT-PROJECT tool so that all options (and uncertainties) were considered in a common, structured format. Over 70 studies were performed during a four-week period. Their conclusions are summarized below:

Study	Current policy	Recommendation	Business impact
SAS reactors: Coil guide beams	Two-yearly inspection/repairs, revealing some deterioration and one case of total failure. New design (carbon steel) has been installed on one reactor and is being evaluated.	Change to new design, carbon steel or stainless steel to allowing four-yearly inspection. New design (carbon steel) shows R4000/month net benefit, but S/S is worth the additional R100 k capital cost to double this to (R8000/month).	**R96,000/year/ reactor** for changing to stainless steel. Payback period: one year compared to current, two years compared to new carbon steel design.
SAS reactors: vessel nozzles above catalyst level	Two-yearly inspection, some with internal stainless cladding.	Lagging to raise temperature and slow corrosion (better payback than cladding) but risk resulting from ineffective insulation.	**R120,000/year/ reactor** risk reduction (two yearly current vs four-yearly with lagging, provided this is effective).
SAS reactors: Plenum/shell weld	Inspect two-yearly, some cracking detected.	Inspect four-yearly, additional risk exposure acceptable and is built into total shutdown cost/ risk model.	− R1200/year (small additional risk).

Study	Current policy	Recommendation	Business impact
Hot quench tower: spray nozzles & shell integrity	Inspect two-yearly; cracking of three spray nozzles found recently. Erosion/corrosion found on shell internal surface.	New localized and focused inspection methods, selective additional spray cladding with stainless steel. Weld overlay too expensive, and does not justify total shell cladding (localized problem).	Selective spray cladding worthwhile (net benefit **c.R170,000/ year/HQT**). Some additional risks in four-yearly vs two yearly (R40,000 to 100,000/year).
Hot quench tower & separator drum: pilot operated relief valves	Overhaul two-yearly; problems with seals and pilot reliability.	Install second pilot unit on each valve, allowing on-line maintenance of pilot and four-yearly on main valve. Also evaluated new seal material, but dual pilot gave better payback.	**R3.4 million/ year** for the 24 × '015' valves. Payback for capital costs is two months (015 valves) and four months (024 valves).
Pumparound system: valves & pump seals	Overhaul two-yearly; isolation valves failure to close prevents on-line cooler cleaning and pump/seal maintenance.	Alternative design (ball valve) is NOT worthwhile, additional risk of four-yearly maintenance built into total shutdown cost/risk model. Other design/ filtering/cleaning options should be explored. Double block and bleed on cooler.	− R40,000/year/ cooling train (additional risk, given that proposed design change is not justified).
Pumparound cooler	Three- to six-monthly cleaning, if isolation valves functional, otherwise two-yearly cleaning.	Three- to six-monthly cleaning: even if short shutdown is required to isolate. Urgently investigate options for filtration, reduced fouling or on-line cleaning. Pumps – replace vessel valves with gate valves when current valves need replacement.	Penalty for continuing to next opportunity can be up to R1 million per month. Worth further study.

(*Continued*)

Study	Current policy	Recommendation	Business impact
Trim cooler	Opportunity-based cleaning, with large impact of fouling upon separator drum temperature and performance.	Seven alternative design configurations and cleaning options were evaluated: best is on-line cleaning (need to check waterbox implications), or valves/pipes for waterside bypass during cleaning. Stainless steel water boxes if shutdown duration impacted waterbox repair and painting.	**R4 million/year** benefit for best solution. Payback for installing bypass or on-line cleaning is two to three months.
Remainder of Synthol trains	Two-yearly planned maintenance plus on-line minor defects and breakdown repairs.	Four-yearly planned work, plus slightly higher rate of minor repairs and short downtime events, built into total shutdown cost/risk model.	Approximate cost/risk neutral (number of events, pm costs, inspection/repair costs).
Planned shutdowns	Two-yearly major Turnaround (15 days), plus 2.5 events per month of avge four hours' downtime (unplanned).	Four-yearly Turnaround, with slightly higher frequency of minor stoppages in second half of cycle.	**R7–14 million per year** average production increase (8 m and 10.7 m SAS trains respectively).
TOTAL SYNTHOL TURN-AROUND	Two-yearly, split into two factories on alternate years.	Four-yearly, split into phases, one per year. For 4 × 8 m SAS and 4 × 10.7 m SAS reactors.	**R40 million/ year net improvement**.

Combining individual tasks into shared downtime opportunities

The above studies were performed individually, and then aggregated manually to explore the combined effect (sharing shutdown opportunities and overhead costs etc.). This process is now automated, to explore all feasible combinations of short, frequent, 'pit-stop' work and larger, longer-interval, major shutdowns. It uses artificial intelligence (genetic algorithms) to experiment with different work bundles, calculating the total cost/risk compromise for bringing some tasks earlier than optimal, and delaying others to take advantage of shared downtime opportunities. The APT-SCHEDULE (see Figure 21.8) tool performs this 'what if?' task extremely quickly, allowing the user to move individual tasks, add, constrain or eliminate work and instantly recalculate the entire work programme in present day $$ values.

Comment from the team involved

'This study has caused a paradigm shift in the way that I think in managing the assets.'

Wiid Venter

Figure 21.8 APT-SCHEDULE: exploration of different task groupings and alignments

Implementation of the results

The shutdown study is now being progressively implemented, with 10 of the 13 recommendations implemented (of the other three, one can only be implemented at next shutdown, the second revealed further opportunities to be explored, and the third proved too costly at the engineering stage). The next shutdown has been deferred to September 2004. In hard numbers, some of the benefits are already being achieved (see Figures 21.9 and 21.10).

Ongoing progress and further studies

In the last couple of years, considerable additional work has been done at the individual problem and decision level. Between July 2001 and April 2002, a further 88 APT studies were conducted, to the value of **R88,06 million**. Individual cases ranging from R14 million to R30,000 in impact.

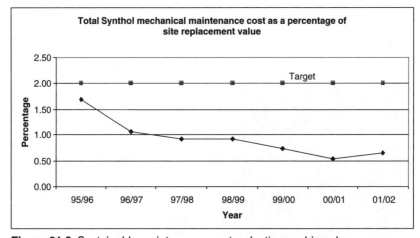

Figure 21.9 Sustainable maintenance cost reductions achieved

Figure 21.10 Production uptime (actual and projected)

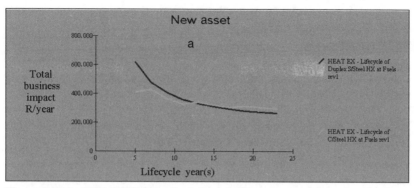

Figure 21.11 APT-LIFESPAN calculation of different materials and optimal renewal policies

In the next 12-month period (May 2002–July 2003), a further 100 studies were conducted to a value of **R304 million** (with individual cases ranging from R20,000 to R50 million impact).

Typical studies in this period have considered the materials used for heat exchangers (lifecycle analysis of various steel types), renewal of major equipment items, and further maintenance and inspection policies. In particular, with the acquisition of site-wide licences in the entire APT suite (seven modules), the APT-LIFESPAN tool (see Figure 21.11) has enabled optimal timing and cost/risk justification of capital projects. These studies have revealed that 'generic' decisions (such as selection of heat exchanger bundle material) cannot be made safely: the evaluation must take account of local usage, criticality and accessibility to influence the choice of materials, the use of bypass options for cleaning etc.

In general, about 80 per cent of the studies have been approved for taking forward – the balance has encountered barriers such as the results and implications have not been fully accepted by all parties yet (lack of training and understanding). Senior management, however, now insist that *only projects evaluated as worthwhile by an APT study will be considered for approval.*

Taking a rough cut of 80 per cent of the anticipated benefits being achieved, this amounts to around **R320 million (approx. US$42 million) impact** in addition to the Synthol shutdown strategy **(R50 million or US$6 million/year)**.

Acknowledgements

The level of support, interest and enthusiasm received from the Synthol staff was vital to the success of the Synthol shutdown study. We are very grateful for the considerable contributions made by Saul Dos Santos, Deon Ludick,

Wiid Venter, Johan Douglas, Riaan Strauss, Graham Duvenhage, Eric Kamfer, Chris Roets, Witold Binkowski, Francois Lombard, Nico Verwey, Philip Smal, Daan Diepraam, Frank Terblanche, Morne van Deventer, Harry Fourie, Kenny Breodenkamp, Johan van der Walt, Coenie Oosthuizen, Jan Pelser and Andries Kinglsey. Thanks must also go to the team and area leaders who made their staff so freely available.

In subsequent studies, others in the Reliability Group, Gasification division and MSM team have 'taken up the torch' and now lead the continuous improvement process – and there is plenty still to do!

22

Summing up – a reality check

The model of excellence can be a powerful factor of change but it is, after all, a tool and certain factors must be present for it to work effectively. I feel I must reiterate these and offer them as presuppositions to be held in mind when attempting change using the model. They are:

1. A model is a model

No matter how powerful a model might appear to be, we must always remember that it is a human construct. It was derived from the experience of many professional managers and engineers but it is not scripture, it is not absolute truth.

The Model of Excellence is a working hypothesis for the best way to plan and manage Turnarounds given our current state of knowledge.

Obviously, based on the work I have done in the last six years, I believe that the Model of Excellence is the best working hypothesis available and I have seen nothing to match it in all my travels, but that is my own, probably biased, view. However, other people who have used it have been very complimentary about the model and this increases my confidence that the model is indeed a good working hypothesis.

The implication of the model being a working hypothesis is that the model itself can be continuously improved. In fact, the Model of Excellence shown in Chapter 14 is the ninth incarnation of the model. Each time it is used, the people who use it are asked to evaluate the model. Based on their responses, the model is regularly updated and, hopefully, improved.

My eyes were opened to the idea of the 'current working hypothesis' by the outcome of a seminar that I attended some years ago. The theme of the seminar was encapsulated in the question:

'What makes an excellent Turnaround planner?'

There were many responses to this question which could be described as 'technical' (e.g. the best way to lay out a job plan, how to arrive at the most accurate estimate for a job duration etc.). There was another class of responses that I found even more intriguing and that I can only classify as the excellent planner's attitude. Over the intervening years I have come to regard these as equally important, if not more important, than technical skills.

The following list shows the presuppositions that excellent planners say they have installed in their minds that guide their thinking when they come to consider the planning of an individual job or a total event. They are designed to knock us out of our comfort zones. They are in fact a working hypothesis. The presuppositions are:

1. You only know how strong something is if you push it beyond its limits.
2. If your grasp does not exceed your reach you'll never stretch.
3. If you meet your duration targets then your planning was slack.
4. If you over-run your on duration then your execution was faulty.
5. Every task on a Turnaround can be done quicker.
6. Every critical path can be made shorter.
7. There is always a better way to do any job.
8. People can always improve their performance.

Note the double whammy provided by combining number 3 and number 4. It is a good recipe for eliminating complacency.

The presuppositions provide what we now call the 'challenge' approach to planning and without them, or something like them, I do not see how it is possible to become an excellent planner. Even if the presuppositions are not true, they are very useful.

In that spirit the Model of Excellence is continuously used, assessed, evaluated and improved as we discover better ways of not only doing things but evaluating them as well.

2. Training

The model is a tool and, like all tools, if you want to use it effectively you need to be trained how to use it. Self-training (using this book) will provide one level of proficiency but professional training in the use of the tool will provide a much higher level. It depends on what level of proficiency is required.

Taking this tool and applying it to your own Turnaround approach will definitely improve the approach, but the question is, is that enough? I have heard it said that the greatest barrier to excellence is adequacy. If we feel we are adequate we never strive for excellence. If what you require is to be adequate then this book should suffice. If, on the other hand, you aspire to excellence then training is essential.

3. Repackaging the obvious

I despair when I listen to some modern consultants and salesmen of maintenance 'systems' talking like snake oil salesmen.

The cry of the nineteenth century snake oil salesmen in the Wild West:

'Drink this secret recipe drawn from the venom of the jibba jabba snake (not to mention a liberal dose of laudanum and alcohol) and it will cure all known ills from dandruff right down to ingrown toenails – and it only costs five dollars a bottle!'

is echoed by the cries of the modern salesmen:

'Use this Marvo maintenance system based on a brand new state-of-the-art approach (not to mention a liberal dose of the blindingly obvious repackaged to look new) that has been developed by our experts and your maintenance strategy will be transformed – your backlog will disappear and your plant will run like a sewing machine – and it only costs half a million dollars!'

The implication is that your present system is no good and you can only improve if you replace your system with their system (with all the blood, sweat and tears that implies).

The Model of Excellence does not make that kind of promise nor does it demand that level of change. Everything in the Model of Excellence *is* obvious, there is no rocket science. Its main strength lies not in what it contains or how it is packaged but in how it is used.

Think of it as a map of an ideal system which if overlaid as a template on your system will tell you where your strengths and the weaknesses lie.

Even if you are dissatisfied with the output of your present Turnaround approach, it will still have some good elements that require no improvements (typically somewhere around 70 per cent!), mediocre elements that will suffice but could be made better with small changes, unsatisfactory elements that require extensive change and finally maybe some missing elements that need to be introduced to the approach. The model will identify each of those elements and provide you with options for improvement based on your specific operating context and circumstances. Therefore you can gain substantial benefits without traumatic change.

It's as simple as that. No earth-shaking transformations. Think of it as polishing up your existing system instead of throwing it away and replacing it with someone else's.

4. The fusion of knowledge

There is an old joke about the definition of a consultant being someone who borrows your watch to tell you what time it is and then charges you for the service (and probably keeps the watch!).

There is some truth in this because some consultants present an improvement system as 'their' system that 'they' will implement but clients find that the implementation of the system requires their own staff to provide lots of 'local knowledge' and do lots of work. Some consultants tend to downplay just how crucial the client's contribution of local knowledge is. Worse still, there are those consultants who seem to totally ignore local knowledge, no doubt judging it to be of little or no use.

The truth is that any successful improvement plan requires the fusion of the consultant's expertise at implementing change with the client's local knowledge. The Model of Excellence is not used as a weapon to impose externally driven change but as a tool which allows the client to make changes internally.

The principles of the model are generic – that is why it can be used in any industry – it is only when the model is implemented that it becomes specific due to the application of the client's local knowledge.

Any consultant with integrity should advise the client that it will be the client's people who will make the changes because they already have the vast majority of the resources and knowledge they need (in many cases they just don't realize that they already 'know').

The consultant's role is to facilitate the change by drawing out that knowledge and then assisting the client to reorganize it in a more effective way to bring about improvement and, in the rare case where it is necessary, add some new knowledge.

5. Excellence

Excellence is not something we achieve, it is something we strive for. In personal terms excellence is not about having good working systems and routines (although these are necessary and important), it is about having an attitude of mind that drives us to continuously improve our performance.

Tony Robbins, an American motivational speaker and NLP Trainer, coined the acronym 'CANI' for Continuous And Neverending Improvement. I have used and enlarged this idea to create my personnel motivator equation whenever I am faced with a problem or issue. It is:

$$CAN I? + CANI = CAN I!$$

The pursuit of excellence in Turnarounds is fulfilled by continuous and neverending assessment of Turnaround performance. The Model of Excellence provides an ideal tool for this. And remember, for best effect, the tool itself should be continuously assessed and improved.

6. A final challenge

One of the themes of this book has been about challenging the way we do things. My final challenge to you, the reader, is this. If you have read thus far, don't stop here. Don't believe or take on trust anything I have said in this book. Prove it in the real world. Take the Model of Excellence and try it out. Benchmark your own Turnaround approach – I can almost guarantee that you will be surprised by what you find.

I always welcome comments and queries regarding Turnarounds, shutdowns and outages. If you feel moved to make comment or have an enquiry, you can reach me at tomlenahan@hotmail.com

Appendix
APT-SCHEDULE:
Optimizing Shutdowns &
Maintenance/Inspection
Schedules

- How do you choose the right time to shut down a plant or process for maintenance, construction or other work?
- What is the best interval for combinations of tasks that share access, transport or other 'per occasion' and overhead costs?
- Which jobs should be done together in the first place?
- How could you investigate alternative work schedules, working practices, seasonal effects or risk exposures?

Selection of the right equipment, the right time, and the right work scope for inspection and maintenance tasks is fundamental to maximizing plant performance and integrity. APT-SCHEDULE identifies the optimal combinations and the cost/risk/performance impact of work programmes. Combined with existing technical assessment and inspection methodologies, it enables operators to develop and demonstrate the best value task combinations and timings saving £millions in lost production, risk effects and overhead costs.

Unique opportunity

APT-SCHEDULE offers a unique opportunity to provide a comprehensive and fully quantified analysis of optimal shutdown and work schedule strategy. This includes risk-based evaluation of all possible intervals, alignments and scheduling efficiencies.

A product of the European collaborative MACRO project (see Asset Performance Toolkit below) APT-SCHEDULE provides sophisticated and rapid 'what if?' analysis of different task combinations, the risks, cost or possible impact of delaying or accelerating individual tasks, and the cost or downtime advantages of grouping them together.

Combined with other MACRO best practice guidelines, this now offers a total programme optimization, based on asset whole life costs, reliability, performance, legal, environmental and safety compliance, and the 'intangibles' of public/customer image, employee morale etc.

When applied to a critical electricity transmission circuit, for example, these analysis methods yielded **28 per cent improvement in system availability at the same time as a 22 per cent reduction in total costs**. These figures are compared to the best achieved by professional work planners.

Used to evaluate alternative shutdown cycles for a process plant, it revealed over **£4 million/year** net improvement. By evaluating the cost/risk impact of longer run lengths, it also identified the bottlenecks and the design adjustments that would allow these savings to be extended by a further £1 million/year.

The analysis process

The APT-SCHEDULE process employs a criticality assessment of key processes, failure modes and effects. This leads to an appropriate level of detail in modelling of deterioration mechanisms (reliability, efficiency and lifespan effects). Individual maintenance and other remedial tasks are identified using a customized RCM and RBI process, a proprietary fast-track screening for low criticality items, and fully quantified cost/risk optimization techniques for those with high criticality.

The 'technically appropriate' tasks are then evaluated for variations in cost/benefit under different timing intervals using APT-MAINTENANCE and APT-INSPECTION tools. Finally, APT-SCHEDULE uses an extremely fast and advanced simulation technique to examine the pros and cons of different alignment options:

- Shared access advantages, downtime opportunities, overheads and logistics cost of combining visits or jobs.
- The risk and performance impact of delayed tasks beyond their optimal timing.
- The additional cost of deliberate 'early maintenance' in some cases to provide optimum task bundling.

The result is a totally *auditable schedule* of the *right tasks* in the *right combinations* at the *optimum intervals*.

Individual tasks evaluated & optimized

Tasks alignments, groupings and visit multiples evaluated

Intelligent auto-searching: genetic algorithms

Optimal total programme, with manual 'what if?' evaluation

Combining hard data with local knowledge

The methodology has been developed to make use of whatever information is already available, i.e. a combination of hard data and engineering, maintenance and operating knowledge.

Range estimating techniques and rapid sensitivity testing mean that opinions, instincts and personal experience can all be used to supplement any hard data that is available. Critical influences are easily identified, and therefore data collection efforts can be focused upon these decision drivers (and not on data that has little or no effect upon the results).

The analysis often reveals that current data collection is misdirected, and the future priorities should be directed at specific and critical features.

Software requirements

APT-SCHEDULE is a PC or network-based software tool. Analyses are stored in an MS ACCESS™ SQL Server or Oracle™ database.

While the software itself is simple to understand and operate, a training course in the underlying principles of cost/performance/risk evaluation is available and recommended.

The Asset Performance Toolkit

APT-SCHEDULE is just one of a range of unique evaluation tools developed by the European EUREKA 'MACRO' project. This is a multi-industry £2 million joint venture, supported by the UK Department of Trade and Industry, addressing the cost/risk evaluation of engineering and management decisions. Other modules of the **Asset Performance Toolkit** include:

APT-MAINTENANCE Cost/risk evaluation of planned maintenance, optimal intervals, preventive, predictive or reactive strategies.

APT-INSPECTION Optimal condition monitoring strategy, including inspection and test intervals, sensitivity and alarm or condition reaction points. Puts numbers to risk-based inspection methods.

APT-PROJECT Cost/risk evaluation of projects, change proposals, modifications, new ideas and other 'one-off' investments.

APT-STOCK Materials and spares strategies, min/max stock, re-order quantities, buffer storage of intermediates, supplier comparisons, stock 'pooling' options.

APT-LIFESPAN Lifecycle analysis, asset replacement timing, repair vs replacement, life extension options, alternative designs, Capex/Opex combinations.

Asset Performance Tools Ltd.
19 Prince Henry House, Kingsclere Business Park, Kingsclere, Hampshire RG20 4SW United Kingdom

Tel. +44 (0)1635 299200   Fax. +44 (0)1635 299555
Email: enquiries@aptools.co.uk   Website: www.aptools.co.uk

Index